The Dark Side
of the Universe

The Dark Side of the Universe

A Scientist Explores the Mysteries of the Cosmos

by James Trefil

Illustrations by Judith Peatross

ANCHOR BOOKS
DOUBLEDAY
NEW YORK LONDON TORONTO SYDNEY AUCKLAND

AN ANCHOR BOOK
PUBLISHED BY DOUBLEDAY
a division of Bantam Doubleday Dell Publishing Group, Inc.
666 Fifth Avenue, New York, New York 10103

ANCHOR BOOKS, DOUBLEDAY, and the portrayal of an anchor
are trademarks of Doubleday, a division of Bantam Doubleday
Dell Publishing Group, Inc.

This book was originally published in hardcover by Charles Scribner's
Sons in 1988. The Anchor Books edition is
published by arrangement with Charles Scribner's Sons.

Library of Congress Cataloging-in-Publication Data
Trefil, James S., 1938–
    The dark side of the universe : a scientist
explores the mysteries of the cosmos / by
James Trefil. — 1st Anchor Books ed.
        p.   cm.
    Includes bibliographical references and index.
    ISBN 0-385-26212-4
    1. Astronomy—Philosophy.   2. Cosmology.
    I. Title.
[QB14.5.T74   1989]   89-33825
523.1′01—dc20   CIP

Contents

The Dark Side
of the Universe

Prologue

CONSIDER, if you will, the thoughts of Archytas of Tarentum, philosopher, soldier, musician, friend of Plato, and follower of Pythagoras.

"Is the universe finite or infinite?" he was asked.

"Suppose it is finite, that it has an ending," he replied. "Then you could walk to the edge of the universe carrying your spear. If you stood at the edge and hurled the spear outward as far as you could, what would happen?

"There is nothing in the void to turn the spear around, so it would move on until it landed. But the spot where it landed would be beyond the point which you said was the edge. You could walk to that farther spot and throw the spear again, then walk to the new landing place. No matter where you say the edge is, there is always a place beyond to which you could throw the spear. With each throw your universe becomes larger. We conclude that the universe cannot have an edge, and must therefore be infinite."

Archytas and his spear provide a compelling image of one of the greatest endeavors ever undertaken by the human mind — the quest to learn the true size and structure of the universe. Unlike many similar searches in the sciences, this quest is motivated almost entirely by the deep-seated curiosity, the thirst for knowledge, that characterizes the human mind. Exploring the bounds of the universe billions of light-years from the earth is unlikely to lead to material gain for the searcher. It will not feed the hungry, nor will it fuel the engines of war. Nevertheless, throughout recorded history, many of the best thinkers that our race has produced have applied themselves to this question, and we who have inherited the fruits of their labors are grateful to them, even if their contemporaries sometimes were not.

We gained our present knowledge of the universe somewhat as Archytas' imaginary spearman might have, step by step. For much of recorded history, the universe that existed in the human mind didn't extend much farther than the blue sky itself, and everyone knew that the sky was held up by a giant (or dragon, or what-have-you). The arguments of men like Archytas were ignored, for it was comforting to believe that we already knew most of what was to be known about the universe. But then the spearman, in the guise of a Polish cleric named Nicolaus Copernicus, threw his spear and the universe became much larger, much more empty, than his predecessors could have believed. In our century the spearman took on the form of the American astronomer Edwin Hubble, who showed us that the stars we see at night belong to just one of many billions of galaxies — galaxies that inhabit a universe Copernicus never imagined.

Today the spear of Archytas is no longer a man-made object, but a quasar located at the very edge of detectability, fleeing away from us at close to the speed of light. We can no longer believe in the simple notion that our "spear" will land somewhere. Even if we supposed it could, the fact would do us no good, since billions of years would pass before it came to "earth." Instead, we study the way the spear moves, we look at the ways the spears are arranged in the sky, and we try to puzzle out the way the universe is put together.

And what a universe it is! Mighty streams of galaxies rush through empty space. Lacy bubbles and voids appear everywhere, mocking those who try to find a simple uniformity in nature. Even the fabric of the universe is not what we thought. At least 90 percent of what is out there consists of material whose form and composition are unknown to us. Hardly a month goes by without some new and unexpected facet of the universe coming to light. As we get closer to the ultimate questions, the rate at which the universe is rendering up her secrets seems to be increasing.

It turns out that most of the universe is invisible to us, giving off neither light nor radio waves to tell us of its presence. It may be that the vast starry dome of the heavens has no more to do with the real working of things than a twig being carried along in a stream has to do with the way the water flows. We may, in other words, live in a universe in which the behavior of familiar forms of matter, such as the sun and the Milky Way, is completely determined by stuff we cannot see, but which we call "dark matter."

As so often happens when new ideas break loose in a science, connections between new ideas and old problems turn up. It has always been difficult for astronomers to explain why stars are clumped into galaxies instead of being spread out more uniformly in space. It seems that the more we learn about the basic laws of nature, the more those laws seem to tell us that the visible matter — the stuff we *can* see — shouldn't be arranged the way it is. There shouldn't be galaxies out there at all, and even if there are galaxies, they shouldn't be grouped together the way they are.

Astronomers peering out into the universe with ever more powerful instruments have seen strange forms taking shape before their eyes. First they saw other galaxies like the Milky Way, then they saw that these galaxies were grouped into clusters. Recently, it has been found that clusters themselves are grouped into long, stringlike structures called superclusters. The most startling (and the most recent) discovery is that between these superclusters are things called voids — huge regions where no stars burn and no galaxies form.

All up and down this great chain of structure, from things inside the Milky Way to the largest supercluster known, we find the imprint of dark matter like footprints in the sand. In the past few years we have come to realize that these two problems — the problem of structure and the problem of dark matter — are connected. We are also beginning to see hints and suggestions that they are also connected to a third important problem—the problem of the origin and evolution of the universe. We seem to have gotten ourselves into a situation, in other words, where our failure to solve a series of problems has led us to the realization that all the problems have to be solved together. A piecemeal approach just won't do.

What I'd like to do in this book is to introduce you to the strange corner of the scientific world where solutions to these sorts of problems are sought. It's a place where theorists throw around galaxies of a billion suns the way a kid throws around marbles, where one discovery barely has time to make it into the headlines before it is shouldered aside by another even more startling. It's a world that stretches the limits of the human mind, a world where quark nuggets, shadow universes, and cosmic strings populate the theoretical landscape. It's a bubbling, heady place where the ferment of new ideas is as exciting and vital as it is possible for a science to be.

We are lucky, because what we are seeing today is a snapshot, a stop-action photo of a new science in the act of being born. Because all the answers aren't in yet, we can concentrate on the process by which scientists move toward certainty, rather than on the certainties themselves. We'll learn enough about how bad ideas get eliminated in science so that I won't feel guilty telling you about my own sentimental favorite in the dark-matter sweepstakes: cosmic strings. As the name implies, these are supposed to be long, one-dimensional cords of dark matter. Unimaginably dense, they were formed when the universe was a fraction of a second old. Later, they served as nuclei around which visible matter collected, and today some theorists suggest that they are to be found in the superclusters that stretch across the sky. If this is so, then the universe is truly stranger than anything we have been able to imagine up to now. It would be possible (in principle) to ride a spaceship to one part of a cosmic string, get out, and walk for a billion light-years—about a tenth of the way across the universe.

Today, two millennia after Archytas first made his argument about the nature of the universe, we stand on the verge of being able to supply an answer to his questions about its size and structure. In huge accelerator laboratories, in distant astronomical observatories, and in giant number-crunching computer installations scientists are starting to close in on the spearman, to narrow his options and constrain his movement. It may be, in fact, that it will be our generation that will have the privilege of providing the final answer to questions that have vexed the mind of man since the dawn of recorded history.

So I'd like you to imagine leaving your comfortable armchair and journeying with me to the outer limits of human knowledge and imagination. Our goal: nothing less than an understanding of the origin, structure, and fate of the universe.

ONE

Expanding Horizons, Shrinking Earth

When I, sitting, heard the astronomer
where he lectured with much applause in
the lecture-room,
How soon, unaccountable I became
tired and sick,
Till rising and gliding out I wander'd
off by myself
In the mystical moist night-air, and from
time to time
Look'd up in perfect silence at the stars.

—WALT WHITMAN,
"When I Heard the Learn'd Astronomer"

EVERY CIVILIZATION GETS the universe it deserves.
I don't mean that the universe actually changes when
our ideas about it change; only an ivory-tower philosopher would make a claim like that. What I mean is that as we learn more about the universe, the questions we ask and the role we assign to the structure of the heavens change.

Everyone starts with the same basic facts—the sun rises in the east and sets in the west, the stars remain fixed in relation to each other, the planets move. What we make of those facts, the sort of universe we construct to explain them, depends on the amount of information we have. The more facts there are, the less freedom there is for the imagination. It also depends on the sorts of things we are willing to accept as valid explanations of what we see. To the Greek colleagues of Archytas, for example, the thought that the earth was not the center of the uni-

7

verse would have been simply unthinkable. To us, such a thought is almost second nature, and this in turn has an effect on the kinds of models of the universe we construct in our minds.

If there is a single lesson that grows out of the progress the human race has made in its successive conceptions of the universe, it is this: the more we learn, the less central our own planet and the human race seem to be. We have come to see ourselves as inhabitants of a small rock orbiting a very ordinary sun in a very undistinguished sort of galaxy. We have also come to realize that things in the cosmos do not happen at random, but that every event is governed by one of a small number of natural laws—laws that we can discover in our laboratories. Everything we see in the sky, like everything on earth, happens in a rational, orderly way. This is our universe, the one we learn about in school, but it is by no means the only universe which the human mind is able to imagine.

Let me give you some examples to illustrate what I mean by this. The oldest written account we have of creation is the Babylonian epic called the *Enuma Elish*. The name derives from the first two words in the epic, which translate roughly as "When on high." Like all creation stories, it provides a coherent, self-consistent account of how the universe came into existence and why it is as we find it. The central act in the creation story is a battle between the leader of the gods, Marduk, and the monster Tiamat, who represents the forces of chaos and is the mother of the gods as well. Marduk proves victorious in this battle and cuts the body of Tiamat in half, using one part to create the earth and the other to create the sky. Later, the gods put stars into the sky to remind humans of their religious duties.

Or again, turn to Egypt. The passage of the sun across the sky is an event of central importance to all people. In most of the early universes, this passage was explained by the movement of some sort of sun god in a chariot. In one form of the story common in the Middle Kingdom in Egypt, the sun god drove across the sky each day. Each evening he would descend into the underworld, where he would do battle with the King of Darkness, fighting his way back to the east so that he could rise again. The red colors at sunset and dawn resulted from the blood shed in these battles.

If you believe that this is the explanation for the sunrise, then you naturally accept the possibility that one night the King of Darkness might win. To an Egyptian, the old question "Will the sun rise tomorrow?" that first-year philosophy students love to argue about was not an academic exercise. A thoughtful Egyptian could not accept the rising of the sun as an automatic event, something to be taken for granted. Each rising was a separate event, a separate miracle, depending on the fortunes of the sun god in the underworld the night before.

For the Babylonians, even the existence of the universe was a contingent fact. We are here because Marduk won his battle with the monster. Had he not done so, primeval chaos would still prevail today. There would be no earth and no heavens and, of course, no human beings to wonder about creation. In both these examples, the important features of the world depend on events to which no immutable laws apply. The universe could be controlled only by the gods, and the gods could be induced to attend to human needs only through the use of ritual.

I suspect that the universes of the spirits and the gods provided much more emotional gratification to those who believed in them than our own universe does to us. It was, after all, a universe in which things happened in a very human way. The attraction of these old beliefs has not totally disappeared even today. A large part of the counterculture movement of the 1960s involved a rejection of the rational, scientific culture of modern America and a return to a more mythical view of the universe. Nonetheless, as emotionally congenial as the old ways were, they left a lot to be desired in the intellectual sense. Battle in the underworld or no, the sun does come up every morning. The motions of the stars and planets may depend on the whim of the gods, but they are regular and predictable. Somehow the juxtaposition of the very personal, contingent truths of the old universes with the regular behavior of the heavens seems difficult to explain, at least to the mind of the twentieth century.

It was the Greeks who first conceived a universe something like that we conceive today. Their ideas were characterized by a lively skepticism. A generation before Archytas, for example, the historian Herodotus made a tour of Egypt. He was shown a temple in which the priests put out food for the god every night. The food was always gone in the morning, a fact which they presented to Herodotus as proof of the god's existence. "I saw no god," he commented, "but I saw many rats around the base of the statue." It's hard not to like someone who thinks that way!

This questioning frame of mind led the Greeks to a universe that was notably unlike those we have been sampling from earlier times. And so compelling was their work that it remained the accepted view of the heavens until after the Renaissance — almost fifteen hundred years. I wonder whether our own version of the universe will last that long!

The man who is usually thought of as the explicator of Greek astronomy is Claudius Ptolemy, who lived in Alexandria in the second century after Christ. Although he bore the same name as the rulers of Egypt at that time, he was not, as far as we know, connected to royalty. He worked at an institution known as the Museum of Alexandria, which functioned somewhat like a modern government research center and laboratory. Scientists and scholars at the Museum carried on their work and wrote about it without being hampered by other teaching responsibilities. Ptolemy collected the measurements of his Greek and Babylonian predecessors, took some of his own, and built on previous work to produce a model of the universe that explained everything that had been observed.

A sketch of Ptolemy's universe is shown in Fig. 1.1. The earth was at the center. Around the earth, crystal spheres turned, carrying the sun, the moon, and the planets. Each sphere turned at a different rate, which explained the motion of the planets with respect to each other and to the stars. The outermost sphere carried the stars. It turned at a rate of a little more than once a day — the one turn explained why the stars moved across the sky each night, the little extra explained why there are different stars in the sky in winter and in summer. To account for the motion of the planets in detail, it was assumed that these bodies rolled around in little circles (called epicycles) which themselves rolled on the main spheres.

The Ptolemaic universe was built on two unspoken assumptions that dominated Greek thought. The first of these is geocentrism: the doctrine that the earth is at the center of all things.

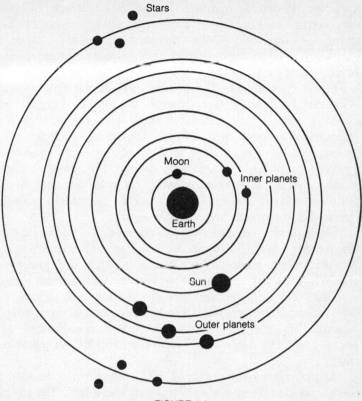

FIGURE 1.1

The second is the idea that motion in the heavens involves circles or spheres. These assumptions are worth remembering, for they are striking examples of the fallacy of relying on ideas that are so obvious that they seem unassailable. Such ideas are unfortunately often wrong. The world just isn't as attuned to the way our minds work as we sometimes think.

Nothing could be more obvious, for example, than that the earth is stationary and that the sun and planets move around it. Anyone who has watched a sunset knows that the sun moves down below the horizon. To believe otherwise requires another vivid and compelling experience to throw doubt on the direct evidence of our senses. The same thing can be said of the assumption of circular motion. The Greek astronomers argued very simply that the heavens must evidently be perfect and unchanging. Consequently, the stars and planets must move in orbits described by the most perfect of geometrical figures. And what is more perfect than a circle?

The circle seems to exert a strange force on the human mind. Whenever I talk about Greek astronomy to one of my classes, I always make a little test. I ask them what the most perfect geometrical figure is. Invariably the answer is the circle or the sphere. I've never heard anyone say a square or a hexagon. If I pursue the point by asking why the circle is perfect, there is usually a pause, and then someone may volunteer that all the points on a circle are equidistant from the center. If I ask why that makes it perfect, there is dead silence.

This sequence, which I've repeated enough times to be sure that it is not a fluke, illustrates the role of unspoken assumptions very well. There is a feel of rightness about such an assumption, and so long as no one raises questions, everything seems logical and true. But the assumption is like the emperor's clothes — once doubt enters people's minds, they suddenly see what's missing. This experience often makes people uncomfortable and angry. I suppose that's why heretics used to be burned at the stake.

The unspoken assumptions of the Greeks formed the core of the scientific universe for a millennium and a half. The people who lived during that period were certainly as smart as we are, yet it never occurred to them to ask questions which to us seem obvious. What does that tell us about the ability of the human race to detect its own unspoken assumptions? Nothing encouraging, I fear.

We would be less than honest if we didn't try to ask similar questions about our own twentieth-century conceptions of the universe. As far as I can see, the major unspoken assumption in twentieth-century cosmology is that there is a rational, mathematically expressible solution for every problem, even the problem of the creation of the universe. Most people react to this statement with incredulity: how can it be otherwise? I imagine a Greek would have reacted the same way had someone questioned the assumption of geocentrism. Yet the Greeks were wrong, and we could be wrong too. In fact, not long ago some people in Berkeley did challenge this unspoken assumption of Western science, claiming that the advent of quantum mechanics would force modern physicists to turn to a view of the universe more in keeping with Buddhism than with traditional thought. As it happened, they had the incredible bad luck to mount their challenge just before the advent of the unified field theories, one of the greatest advances ever made in science. We'll talk about these theories later; for the moment we simply note that they were developed at a time when the very foundations of science were under attack and, thanks to their success, brought an end, at least temporarily, to any serious attempt to challenge our unspoken assumption.

I hasten to add that I believe our assumption is correct; I believe there are rational solutions to the problems of cosmology, and that these solutions can be found by the honest application of the scientific method. But since others have believed just as strongly as we do in their assumptions and were later proved wrong, it behooves us to be aware of the fact that we too proceed on the basis of assumptions that will be validated only when all the answers are in.

If you believe that the earth is at the center of the universe, then your universe can be a relatively small place. If you believe that the earth orbits around the sun, on the other hand, the universe must be a good deal larger. The reason for this state of affairs is an effect called parallax.

To illustrate parallax, try holding your arm out with your finger pointed, then close one eye. With your open eye you'll see your finger line up with some distant object—a mark on a wall, a tree, or whatever. Now close the eye that has been open, open the other, and look at your finger again. You'll notice that it is no longer lined up with the same point, but with something else. This shift, in a nutshell, is parallax. The experiment you have just done is diagrammed on page 14 in Fig. 1.2 (top). When you are looking one-eyed in the direction of your finger, your line of sight is as shown, and you see point A lined up with your finger. When you look through the other eye, your line of sight shifts and you see your finger lined up with point B. There's no mystery here, just some simple geometry.

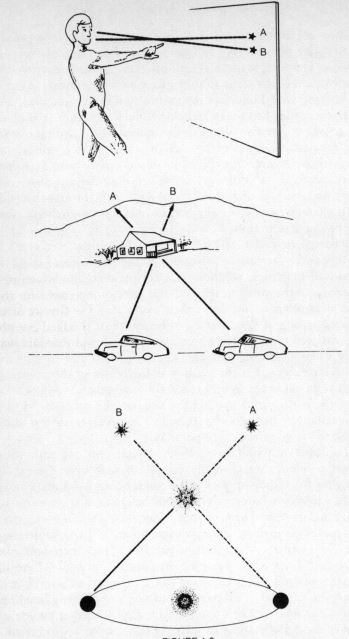

FIGURE 1.2

Examples of this phenomenon are many. When you drive along a highway, as shown in Fig. 1.2 (middle), a house may appear to move against the distant background. The apparent motion is caused by the fact that as your car moves you are looking at the house from different points along the road.

Now suppose that the earth actually does orbit the sun, as shown in Fig. 1.2 (bottom). If you look at a nearby star in the summer, you will see it line up with some more distant star such as the one labeled A. Six months later, when the earth has moved halfway along its orbit, the nearby star will be lined up with something different, such as the star labeled B. Just as a house appears to move when your car goes down a highway, the stars and planets should appear to move as the earth goes around its orbit.

Stellar parallax — the apparent motion of the stars caused by the earth's orbital motion — can be seen with modern telescopes, but the effect is much too small to have been detected with any of the astronomical instruments available to the Greeks or to scientists of the Middle Ages. All they had was naked-eye observation. For them, there was no parallax, and this fact was held to be important evidence against any attempt to take the earth from the center of the universe and put it somewhere else. So although a few Greek scientists such as Pythagoras and Hipparchus suggested that the sun might occupy the central place in the cosmos, their arguments could not be taken seriously. The reason for the apparent lack of parallax — that the universe is just too large for any parallax to be seen with the naked eye — was not suggested until well into the seventeenth century.

The earth-centered universe of Ptolemy was especially congenial to the way medieval scholars thought. Once it was introduced to Europe in the twelfth century (by way of translation from the Arabic texts of the ancient Greeks), it swept through the universities like a storm. The only opposition to the Ptolemaic universe that I have been able to find came from one Etienne Tempier, Bishop of Paris, who in 1277 issued 219 condemnations of the new Greek learning that was being taught in the academies. His main objection seemed to be that by talking about laws of nature, the science faculties were somehow lim-

FIGURE 1.3

iting the power of God. This argument, I confess, leaves me a little cold.

In any case, the Ptolemaic universe was quickly amalgamated with Christian thought, as can be seen in Fig. 1.3, which shows a woodcut from Martin Luther's *Biblia*, published in 1534; in it God is looking down on a series of concentric spheres with the Garden of Eden at the center.

The Ptolemaic universe fitted the preexisting view of a moral universe in which man occupied a middle place, with hell beneath his feet and heaven above. The spheres of the stars and planets were thus between man and heaven. Volcanoes provided glimpses into the underworld, and the blue of the daytime sky was a reflection of the glory of heaven. Demons walked at night, when the heavenly glow was blocked by shadows — further proof of the validity of this cosmology. That a great scholar and theologian like Thomas Aquinas (c. 1224–1274) should establish the right of scientific reason to operate according to its own rules within the larger framework of the Christian faith only strengthened the idea of the earth-centered cosmology; it went hand in hand with the Christian faith.

The medieval universe, then, combined the best of both worlds: the rational, phenomenon-oriented astronomy of the Greeks and the secure and emotionally satisfying spiritual interpretation of life offered from the earliest times by mythology. No wonder church and secular authorities of the late Renaissance were so reluctant to abandon this synthesis.

Yet abandon it they did, for the old Greek assumption embedded in the Ptolemaic system could not stand the test of improved observational techniques. When Ptolemy worked out the sizes and rates of rotation of all his spheres-within-spheres, he adjusted them to match the observations available in his day. This was analogous to setting your watch by the clock at the U.S. Naval Observatory. If you make the adjustment properly, everything will work well for a while, be it universe or watch. But as time goes by, any little imperfection in your watch will take it out of synchronization with the correct time, and the longer you wait, the more apparent the discrepancy will be. The same thing happened to Ptolemy's clocklike universe. By the late Middle Ages, the old clock was showing its imperfections all too clearly. Astronomers would calculate where a planet ought to be and find that it wasn't there. Some tried to tinker a little with the spheres, but so great was the respect for the ancient learning that no one thought seriously of doing a massive overhaul of the system.

No one, that is, except an isolated Polish cleric working in a cathedral on the Vistula River. Nicolaus Copernicus was not bothered by the troubles that the Ptolemaic system was having matching theory with observations; in fact, Copernicus wasn't much of an observational astronomer. Instead of gazing at the stars, he spent his spare time in his study, juggling mathematical figures. His goal: to see if it was possible to construct a universe that worked as well as Ptolemy's, but in which the earth circled the sun, rather than vice versa.

As it happened, the answer to this question was yes, but only because of the poor way in which the Ptolemaic system fitted the data. Today we honor Copernicus not because he produced the modern view of the solar system (he didn't) or because his system was simpler than Ptolemy's (it wasn't), but because he was the first person in "modern" times who had the courage to think the unthinkable and the courage and perseverance to carry his idea beyond the realm of philosophical speculation. It was he who pointed out that the emperor's clothes might be missing, so that after him everyone came to see geocentrism as just an assumption, one that could be challenged like any other.

Once this idea was accepted, the closed medieval universe had to be abandoned. To account for the motions of the planets, it was necessary to put the sun at the center and think of the earth as moving in orbit around it. And to reconcile this finding with the absence of parallax it was necessary to assume that the stars and planets were much farther away than anyone had previously imagined.

Notice how this argument works. We introduced parallax by lining up a distant object and your finger first with one eye, then with the other. Try doing the same thing with two faraway objects, for example, a hill and a cloud. You'll see at once that although alternating eyes will cause both the hill and the cloud to move about with respect to nearby objects, they don't jump around with respect to each other. This is shown on the left in Fig. 1.4.

The reason that parallax disappears is that the difference in angle between the two lines of sight is too small for the eye to detect. To the eye, the two lines of sight are exactly parallel. It makes no difference which line you choose, the result is the same — the objects do not move with respect to each other. In the same way, if the stars and planets are far enough away from the earth (see the right-hand side of Fig. 1.4), no parallax can be seen without advanced detection systems. Thus acceptance of the Copernican universe requires that the stars be very far from the earth. In exchange for a better model of the solar system than Ptolemy's — one that keeps better time — we have to give up the comforting finiteness of the medieval universe and face the fact that we live in a universe which is, to all intents and purposes, infinite in extent, as Archytas and his spearman suggested.

Archytas was not the only one who ever considered the possibility of an infinite universe. In the fourteenth century, he had a counterpart in a cardinal, Nicholas of Cusa, who in his book *Of Learned Ignorance* argued that wherever a man stands he thinks himself to be at the center, so that "the universe has its center everywhere and its circumference nowhere." This amaz-

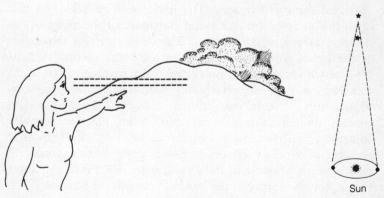

FIGURE 1.4

Sun

19

ing prefiguration of the modern picture of the universe was largely ignored by scholars. Hence when the horizons of the universe were pushed back by Copernicus, hardly anyone was ready for the shock.

When the dust had settled, however, something very much like our modern universe had emerged. Mankind no longer thought of itself as occupying the center, but instead viewed its habitat as a rocky sphere circling the sun, along with the other five planets then known. Outside the orbit of Saturn, at unimaginable distances, the stars swam in the void. The sky was no longer a roof enclosing all creation, and it no longer shone with the comforting radiance of heaven.

But something was gained to compensate us for this loss. Though the sky was no longer a roof, it was at least a doorway —a frontier. If we must admit that we are not the center of the universe, we can take comfort in our ability to face the fact squarely, without flinching, and then go on to wring from our expanded universe its innermost secrets. Perhaps when we have finished doing so, we will reckon the gain well worth the cost.

Some Personal Remarks

The replacement of the emotionally satisfying spiritual universe with a mechanistic, scientific one has not met with universal approval through the ages. The lines by Walt Whitman that open this chapter are as clear a statement of the voice of opposition as one is likely to get. The counterculture movement of the 1960s is but the most recent outbreak of this attitude. No one who teaches in a university can remain ignorant of the fact that many people, even highly educated people, agree completely with the sentiment Whitman expressed. There is a widespread, though seldom articulated, belief that by studying something analytically, we destroy its beauty. "We murder to dissect," as Wordsworth put it. Before going on to examine the details of the universe of the 1980s, therefore, I'd like to argue that it, too, is worthy of the kind of aesthetic admiration Whitman bestowed on his starry night.

In one sense, the point that Whitman makes is unarguable:

direct sensory experience is always more meaningful than analysis. This is one reason why the space program places such a great emphasis on producing photographs of objects being explored, even if the pictures have to be in false colors. But this is not the same as saying we have to see things directly with our own eyes in order to appreciate them. We can enjoy photographs of the Himalayas, derive aesthetic enjoyment from them even if we never travel to Katmandu.

Look at it this way: if you go out and "look in perfect silence at the stars," you'll see at most about twenty-five hundred stars and the odd planet or two. It's impressive enough, I grant you, but it's a drop in the bucket compared to what's out there. Our galaxy alone contains over ten billion stars, and we're only one galaxy among billions. If we restrict our knowledge of the universe to what we can perceive directly, we are deliberately impoverishing ourselves, accepting an experience infinitely less rich than it could be. It is indeed wonderful to see a star-studded sky, but there is beauty in an infrared picture of the galaxy, a photo of the rings of Saturn taken by the *Voyager* space probe, or a computer-generated picture of an airy cluster of galaxies spreading across the sky at unimaginable distances from us. Surely there is room for both the poet and the technologist in modern astronomy.

But there is a deeper sense in which critics of science have failed to grasp what has happened to the modern universe. It's true that we have traded a universe of human scale for one unimaginably larger and more complex. Yet isn't it important that our universe satisfy our intellect as well as our emotions? Isn't it important that as large and complex as our universe appears to be, the human mind can still comprehend it and discover the basic order which underlies the complexity? Yes, we have traded a universe in which humans imagined that they could control the gods through ritual and ceremony for one in which our control over nature comes through a grasp of her basic laws, but would you really trade a surgeon for a shaman if you had appendicitis? We have indeed traded a universe in which God intervened in human affairs for one in which the role of God is to devise the laws of nature and then let things unfold without any need for further interference. But wouldn't you rather have a God who knew how to do things right in the first place? All I can say is that for me, a universe dominated by the hard, crystalline truths of physical law is every bit as beautiful as any universe ever devised by the human mind, and I wouldn't trade the modern universe for anything that has gone before.

TWO

Discovering the Galaxies

The good mate said: "Now we must pray,
For lo! the very stars are gone.
Brave Admiral speak, what shall I say?"
"Why, say, 'Sail on! sail on! and on!' "

— JOAQUIN MILLER, "Columbus"

ONCE THE SUN was moved away from the center of the universe and the limits of the universe were pushed back, people started to ask how the universe was put together. The astronomers of the eighteenth and nineteenth centuries were like voyagers who had discovered a new continent that needed to be explored and charted. Since the ability to see distant objects in the sky was limited by the power and sensitivity of the telescope, the observer using a larger telescope often wound up making a great discovery. As time passed, the survey of the universe brought two central questions into debate: (1) How big is the Milky Way? (2) Are there other "island universes" — galaxies — in the sky?

With each new discovery, the perceived size of the universe increased. Over and over again astronomers found themselves looking at a cosmos that was larger than anything they had ever imagined. The historical progress of the ancient spearman is obviously tied to modern man's ability to deal with the two questions posed above.

If you step outside on any clear summer night and look up, you can hardly miss seeing the Milky Way. Billions of stars (most invisible to the naked eye) come together to form a frothy path through the darkness. It is the most spectacular feature of the night sky. It also provides the first clue to the structure of the universe outside the solar system.

Thomas Wright, an English natural philosopher, is generally regarded as the man who first speculated effectively about the structure (if not the size) of the galaxy we now call the Milky Way. Written around 1750, his work has a mystical, almost medieval ring to it. Regarding the universe as the handiwork of God, he saw the study of it as akin to theology. He reasoned that God would have made the universe perfect; therefore it must be constructed of spheres. The Newtonian astronomers who worked out the elliptical orbits of the planets had driven the sphere out of the solar system, but Wright made a valiant attempt to reinstate it in the larger universe.

Wright's universe is sketched in Fig. 2.1. The stars are contained between two concentric spheres. Human beings, anchored to the planet earth, are located somewhere between the

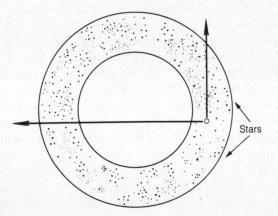

Stars

FIGURE 2.1

spheres, as shown. When we look out in a direction tangent to the spheres, our line of sight will cross many stars. If we look outward or inward along a radius, however, we see very few. This arrangement, according to Wright, accounts for the existence of the Milky Way; it marks the direction tangent to the spheres.

Implicit in this argument is a statement about the size of the universe—that the radius of the inner sphere has to be large enough so that when we look inward through the center of the sphere we do not see a large concentration of stars on the far side. The radius must also be large enough so that when we look out tangentially, we do not see the curvature, and the Milky Way appears more or less as a straight line.

I suspect that Wright never pursued this point in order to estimate the size of the universe because he regarded science as a means of drawing moral lessons, rather than as an end in itself. In 1966 some of his unpublished manuscripts were uncovered, and one, entitled "Second or Singular Thoughts upon the Theory of the Universe," proposed an even more theologically based model, in which the sun was at the center of a star-filled sphere. Using the analogy of earthly events such as the Lisbon earthquake, Wright explained the Milky Way as a kind of celestial lava flow within the sphere. His goal seems to have been a system in which the orders of the physical and moral universes were exact analogues of each other.

It is hard to say whether Wright was a throwback to earlier modes of thought or a precursor of modern science. The social Darwinists of the late nineteenth century also tried to draw an analogy between the biological order of nature and the social order of mankind. Marxists do the same. Nor is this tendency to draw analogies between science and social or moral truths absent from our own, supposedly more enlightened, times. I recall, for example, reading a treatise in which it was argued that the laws of quantum mechanics proved that a particular radical feminist political agenda popular in the late 1970s was the only political structure consistent with the natural world.

But just as those who followed Wright realized that the shape of the universe has nothing to do with the moral teachings of the Anglican Church, and just as we realize that our moral duty to care for the less fortunate has nothing to do with whether or not nature operates according to the survival of the fittest, I hope that our descendants will finally come to see that the laws of nature are value-free, that they contain no lessons whatsoever about how we should organize society or live our lives. In deciding such matters we are on our own.

Not many men of science followed Wright in his search for a moral order in the universe. The first modern attempt to explore the Milky Way was made by William Herschel. Born in Germany in 1738, he began his professional career as an oboist in a military orchestra, emigrated to England, and became a successful musician and instrument maker. He even did a little composing on the side. That is not his chief claim to fame, yet when the Adler Planetarium in Chicago put on a display of old astronomical instruments, they had a tableau of Herschel peering at the sky while one of his harpsichord pieces played on the soundtrack.* Eventually Herschel decided to make his astronomical hobby his main interest, and despite the loss of income involved, he became a professional astronomer.

His idea was to map out the sky by counting the number of stars that he could see when he pointed his telescope in a given direction. Assuming that stars are distributed more or less uniformly in space, he reasoned that when he looked a long way through the universe (or, as we would say, through the galaxy), he would see a great number of stars. If he looked in a direction toward the edge, he wouldn't see so many. The conclusion of his survey: the universe was flat but irregularly shaped — something like a squashed amoeba. A sketch of Herschel's Milky Way is shown in Fig 2.2.

While this stellar surveying was going on, another series of discoveries was being made. These concerned some rather mysterious objects in the sky called nebulae (clouds). When the viewing conditions are right, these objects can be seen with the naked eye. They were known to Arab astronomers in the eighth century, and you can readily see one if you look at the constellation Andromeda, which is high overhead in the autumn and winter sky. What you see is a fuzzy white patch of light,

---

* The music sounded as if it had been written by Handel.

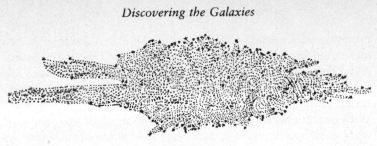

FIGURE 2.2

too big to be a star but not very bright or spectacular. When some of these nebulae were examined with telescopes, it was seen that many of them contained individual stars visible against a luminous but cloudy background. It was the German philosopher Immanuel Kant who, in 1775, first suggested that these nebulae might be other "island universes" like our own. Without a means of making a more detailed examination of the structures, however, the question of the identity of the nebulae remained more philosophical than scientific.

By 1845, William Parsons, the Earl of Rosse, had a telescope built in England whose main working part was a mirror with the then unheard-of diameter of seventy-two inches. With this powerful instrument, Rosse was able to discern a spiral structure in many nebulae in the sky. Although he worked in the days before astronomers took photographs, his sketches of nebulae remind one strongly of modern pictures of galaxies. The familiar flat disks with spiral arms are clearly visible. Because some nebulae seemed to have the flattened shape attributed to the Milky Way, Rosse's work rekindled interest in the old island-universe suggestion.

The argument about the nature of nebulae continued throughout the last half of the nineteenth century and well into our own. Some nebulae showed a spiral structure, while others appeared to be swirling clouds of gas in which only a few irregularly scattered stars were to be found. If nebulae were really distant island universes like our own, then why were some of them so obviously clouds of gas? And if the nebulae were all inside the Milky Way, why did the spiral-shaped ones look so much like large collections of stars?

This debate raged in astronomy for over sixty years—two

professional generations—and, as we shall see, it wasn't completely resolved until the 1920s. The final answer to the question about the nature of the galaxies is "all of the above." There are nebulae that are associated with the Milky Way—gas clouds sprinkled with a few stars—and there are others that are galaxies like our own. There is no single solution to the riddle of what they are, because it turns out that there are nebulae of both kinds.

This is an important piece of history for several reasons. Later in the book we will be considering a similar question, which can be stated as "What is dark matter?" Until recently, astrophysicists who thought about this question tended to be like their predecessors—they attacked the problem with the assumption that there was only one kind of dark matter. They tried to find something that would do all the things that dark matter "ought" to do. I am happy to say that when this quest ran into problems, my colleagues showed sufficient flexibility to begin looking into the "all of the above" possibility.

The long debate over the identity of nebulae exemplifies again how easy, how natural, it is to wear intellectual blinders. Reading through the papers on the debate about the nature of nebulae is a profoundly humbling experience. Here were two generations of the best scientific minds in the world worrying over a problem, making measurements, and arguing, and not one person (at least none that I could find) suggested what seems to be the obvious solution. Yet it is clear from reading their papers that these people were good scientists with first-rate minds, far more talented than I or most of my friends. How could they have overlooked something so obvious for so long?

Depending on your general level of optimism, you can look at such an occurrence in two ways. You can see good scientists overlooking obvious solutions for a long time and from this despair of intellectual progress, or you can reflect that despite being stalled, the scientific process eventually reaches a correct answer.

But whatever conclusion you choose to derive from the great debate over the nature of nebulae, one must admit that it was an extremely important one. It concerns nothing less than the ultimate size and structure of the universe. It also illustrates better than anything I know the intimate connection that exists in science between the quality of the instruments available and the progress that can be made in answering important questions. As we shall see, it was not brilliance that solved the nebula question, but the availability of the hundred-inch telescope on Mount Wilson in California.

The connection between instruments and progress is critical, because to prove that a particular nebula is (or isn't) in the Milky Way one must (1) find out how far away the nebula is, and (2) find the size of the Milky Way. Thus, knowing the nature of nebulae necessarily involves measuring the size of the galaxy and the distances to stars.

The measurement of distances in astronomy is one of those subjects that practitioners prefer not to discuss in polite company. There are too many dirty secrets swept under a variety of rugs. Measuring these distances is not like measuring the length of a table—simply laying a ruler down, marking its end, and laying it down again until you come to the end of the table. In astronomy, the "rulers" that work for nearby stars won't work for stars farther away, so a distance scale must be built up by using a first ruler out to its limits, devising a second ruler to match the first in regions of overlap, and then using the new ruler as far as it will go, then calibrating to yet another ruler to go still farther, and so on. This method seems to work, but it certainly does give the measurement of great distances a certain Rube Goldberg, jury-rigged look.

By the late nineteenth century, there were two "rulers" in use to measure the distances to stars. The simplest is triangulation, illustrated in Fig. 2.3 (page 30). If the angle toward a star is measured when the earth is at position A, and then measured six months later, simple geometry and a knowledge of the diameter of the earth's orbit enable us to calculate the distance to the star. Conceptually simple, this method depends on our ability to measure small differences in angles. When the distance to a star is very large compared to the diameter of the earth's orbit, the angles measured from the two positions become very similar and we lose the ability to differentiate between them. Triangulation can be used to determine distances up to about 150 light-years—a very small fraction of the diameter of the

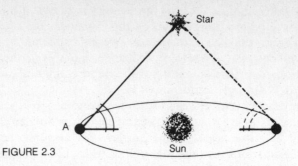

FIGURE 2.3

Milky Way. In the nineteenth century, with poorer telescopes than are now available, the triangulation "ruler" could hardly be used even as far as that, and could certainly not measure distances to the edge of the Milky Way.

Triangulation, then, gives out a short way from the earth. The next ruler, which I won't explain in detail, involves a somewhat more complicated use of geometry together with measurements of the apparent motion of stars. With these techniques, astronomers have developed a ruler for measuring distances out to several hundred light-years. This ruler, too, was available in the late nineteenth century, but it contributed little to estimates of the size of the Milky Way because it didn't reach far enough.

This situation changed in 1908, when Henrietta Swan Leavitt, working at the Harvard Observatory, made an important discovery about the type of stars that astronomers call Cepheid variables.* These stars exhibit a regular pattern of brightening, dimming, and brightening when observed over periods of time from a few weeks to a few months. What Leavitt noticed was that the brighter the star was, the longer the pulsing took. This means that by watching one of these stars, we can determine the period of the pulsation and discover how bright the star is (or, equivalently, how much energy it is pouring into space). If you compare this number with the amount of light we actually receive from the star, you can tell how far away it is. This third

---

* The name arises from the fact that the first such star was observed in the constellation Cepheus, which is named after a mythical king.

"ruler" eventually allowed astronomers to measure distances of over a hundred million light-years. As we shall see, it led eventually to the resolution of the nebula problem.

An interesting sidelight upon the state of astronomy in the early twentieth century is that although it was known that the earth was not the center of the universe, it was believed that the sun *was* at or near the center of the galaxy. This was not an unspoken assumption analogous to Ptolemy's notion about the place of the earth; it was based on such data as were in hand at the time. I find this little fact interesting, because it shows that whenever we can misinterpret the data so as to make our home system seem more central to everything than it is, we leap at the chance. I wonder how much of the current enthusiasm for finding extraterrestrial civilizations like our own is fueled by the same tendency.

The man who finally determined something like the size we now assign to the Milky Way was an interesting American astronomer, Harlow Shapley. Born in Missouri, he began his career as a crime reporter for a small-town newspaper in Kansas. Covering the drunken brawls of oilmen and cowboys apparently became boring, so he decided to go to college. His description of the way he settled on astronomy at the University of Missouri is unusual, to say the least:

"I opened the catalogue of courses . . . the first course offered was a-r-c-h-e-o-l-o-g-y, and I couldn't pronounce it! I turned over a page and saw a-s-t-r-o-n-o-m-y: I could pronounce that, and here I am."

What Shapley discovered can be understood from the drawing in Fig. 2.4 (page 32). We know now that the Milky Way galaxy is a complex structure. In addition to the familiar flattened spiral disk, it is surrounded by a spherical arrangement of clusters of stars, called globular clusters. Containing millions of stars each, these clusters (and the Cepheid variables in them) could be easily detected with the telescopes available to Shapley and his colleagues. When counts of clusters were done, it was found that they tended to be seen only on one side of the sky. Shapley argued that these clusters are actually distributed uniformly around the Milky Way, and showed that the only way we could understand the observations was by (1) making the galaxy much

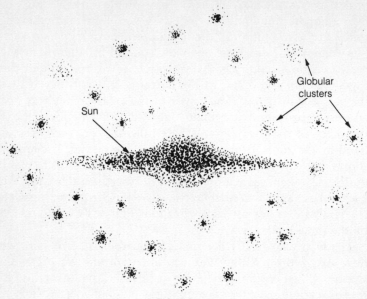

FIGURE 2.4

larger than had previously been thought, and (2) moving the sun away from the center, as shown.

With Shapley, then, we approach the modern view of the galaxy. It is a flattened disk roughly 100,000 light-years across,* and our home system is located to one side, in the low-rent suburbs a third of the way to the rim. Shapley, we might say, played the same role in galactic astronomy that Copernicus played in the solar system — he moved the earth away from the center of things. Thus the last vestige of geocentrism was expunged from science. At the same time, the known universe had expanded to a size that would have seemed beyond belief to men such as Herschel who first tried to map out the stellar world.

The critics of Shapley's conclusions believed that spiral nebulae were, in fact, other galaxies like our own. They reasoned

---

* Shapley's original estimates were somewhat higher than this.

that if the Milky Way was really as large as Shapley said, there seemed little chance that the nebulae could be outside its boundaries. This feeling was supported by the discovery of bright new stars, called novae, in some of the spirals. These stars were so bright that they looked as if they could not be far away. Today we know that they were supernovae, the explosions of giant stars. We also know that for a brief period a supernova, fueled by nuclear reactions, can outshine an entire galaxy. In the 1920s, when nuclear power was not well understood, this explanation would not have carried much weight. The tide seemed to be running against the island-universe theory, and the universe was held to consist of one galaxy, not many.

So confused was the situation that in April 1920, Shapley and Heber Curtis, one of the chief proponents of the island-universe hypothesis, held a debate at the Smithsonian Institution on the question of the structure of the universe. Sponsored by the National Academy of Sciences, this debate is looked on by astronomers as the equivalent of the famous Huxley-Wilberforce debates on the validity of evolution. Shapley presented his evidence for the size of the Milky Way, and Curtis argued in favor of the existence of other galaxies like our own. No one "won" the debate, primarily because the two men discussed different issues. Each was correct in his own domain. The Milky Way is indeed very large, as Shapley argued, but the distances to other galaxies are even larger.

The nature of nebulae was finally ascertained in 1923 when the American astronomer Edwin Hubble became one of the first scientists to get time with the new hundred-inch telescope on Mount Wilson, near Los Angeles. With this instrument Hubble was able to pick out individual stars, including Cepheid variables, in nearby galaxies. Using the correlation between pulsation and brightness developed by Leavitt, Hubble demonstrated that the distances to the spiral nebulae were to be measured in millions of light-years, distances far greater than those assigned by Shapley to the size of the galaxy. Once more, the universe expanded as our ability to see into it improved. Not only were there other island universes, but they were much farther away than anyone had ever imagined anything could be.

The spiral nebulae are indeed star systems like our own, located at enormous distances from us. Other nebulae—those with relatively few stars and lots of wispy material—are gas clouds in our own galaxy. To tell the difference between the two, we needed a telescope capable of determining that one set of nebulae were more distant than the other. Once this was done, the problem was solved.

A Word About Clusters

Before continuing this discussion of Hubble's work, we should note one fact about his island universes. Since his time, it has become clear that galaxies are not distributed randomly in space, but tend to be clumped together in structures called clusters, and that the clusters themselves are gathered into superclusters. Although Hubble had no way of knowing this fact, it turns out that explaining the uneven distribution of galaxies constitutes one of the major problems, some would say *the* major problem, in modern cosmology.

The Expanding Universe

Important as Hubble's proof of the existence of other galaxies was, another discovery he made as part of the same study was even more striking. Looking out at nearby galaxies, Hubble could see that they were moving away from him, and that the farther away the galaxy was, the faster it was moving. This discovery is so astounding—so fraught with implications for modern cosmology—that we should consider the basis of the reasoning on which Hubble made his claim.

When you stand next to a highway and hear a car blowing its horn as it goes by, you notice that the sound of the horn changes when the car passes you. Its pitch is higher as the car approaches, lower as it moves away. This is an example of what is called the Doppler effect. It is explained most easily by reference to Fig. 2.5.

When a stationary car emits a sound wave, as in the top diagram of Fig. 2.5, a series of concentric rings of compressions

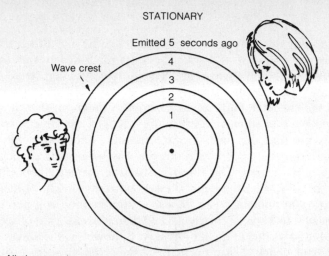

STATIONARY

Emitted 5 seconds ago

Wave crest

All observers hear same wavelength. Distance between crests is equal.

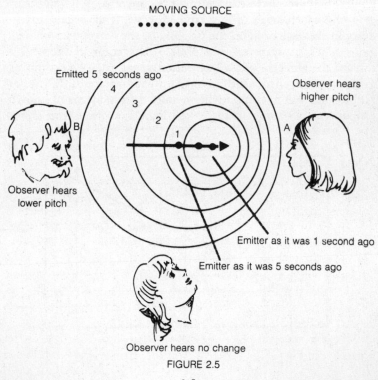

MOVING SOURCE

Emitted 5 seconds ago

Observer hears
higher pitch

B

A

Observer hears
lower pitch

Emitter as it was 1 second ago

Emitter as it was 5 seconds ago

Observer hears no change

FIGURE 2.5

and rarefactions in the air move out from the car. When these strike our ear we hear a sound, and its pitch depends on how closely bunched the waves are. The more waves that strike our ear each second, the higher the pitch.

If the car is moving, as in the bottom diagram, then it will actually travel some small distance between the time it emits one wave and the time it emits another. Each wave will be centered on the spot where the car was when that particular wave was emitted. The result is that the formerly concentric pattern is replaced by the skewed pattern shown. Someone standing at point A when the car is approaching will perceive the waves to be more closely packed than normal. That observer will hear a sound of higher pitch. At B, however, the waves are less closely packed than normal and the observer hears a sound of lower pitch.

This explains the Doppler effect, and it also explains how Hubble discovered the expansion of the universe. What happens to sound can happen to any kind of wave, from ocean surf to light. In the case of light, the packing of the waves as an object approaches is perceived as a shift of the color of the object toward the blue; the spreading out of waves as an object moves away is perceived as a shift toward the red—a "redshift." What Hubble did was to compare light emitted from atoms of known elements in nearby galaxies to the light from the same atoms

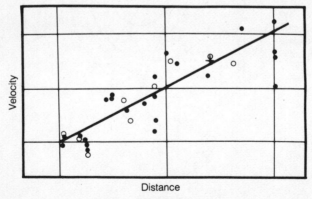

Distance

FIGURE 2.6

emitted in terrestrial laboratories. He found that the light from distant galaxies shifted toward the red end of the spectrum, from which he concluded that the galaxies are moving away from the earth.

Another pattern emerged when he looked at how far away the galaxies were. Some of his original data are sketched in Fig. 2.6. Hubble saw that the data indicated a trend—the farther away the galaxy was, the higher its redshift. I have to admit that this trend is not overwhelmingly evident in the figure (although it is very clear in modern data). Hubble's discovery has to be attributed in part to good experimental technique and in part to an inspired guess as to what would emerge when better measurements were done.

However Hubble perceived the pattern that would eventually emerge from his data, the discovery of the galactic redshift was one of the most profound results ever to come from astronomical observations. It carried within it the seeds of our modern picture of an evolving universe, the picture we call the Big Bang.

THREE

The Big Bang

Why are people born? Why do they die? And why do they spend so much time in between wearing digital watches?

—DOUG ADAMS,
The Hitchhiker's Guide to the Galaxy

THERE'S NO DOUBT about it—if you want to know the answers to the most fundamental questions science can ask, you have to turn to cosmology. Throughout history, cosmologists have taken it upon themselves to answer questions like "How did the universe begin?" or "How is it built?" or "What is its future?" When you ask these questions of a cosmologist today, the answer you get is couched in the language of the accepted model of our time—the so-called Big Bang. This model is a logical outgrowth of the discoveries about galaxies made by Hubble.

If distant galaxies are really receding from the earth, and if more distant galaxies are receding faster than nearby ones, a remarkable picture of the universe emerges. Imagine that the galaxies were raisins scattered throughout a rising lump of bread dough. As the dough expanded, the raisins would be carried farther and farther apart from each other. If you were standing on one of the raisins, how would things look? You wouldn't feel any motion yourself, of course, just as you don't feel the effects of the earth's motion around the sun, but you would notice that your nearest neighbor was moving away from you. This motion would be due to the fact that the dough between you and your nearest neighbor would be expanding, pushing the two of you apart.

If you then looked at a raisin that was twice as far away from you as your nearest neighbor, you would see that one receding as well. It would be moving twice as fast as your neighbor, because there would be twice as much dough between you and it as there was between you and your neighbor. The farther away you looked, the more dough would separate you from the raisin you saw, and the faster it would be moving away from you.

This is, in effect, what Hubble saw when he looked out at the rest of the universe from Mount Wilson. It was a universe dominated by an overall expansion of space itself, the galaxies being carried along like the raisins in dough. It was an expansion in which an observer in any galaxy would consider himself to be standing still, and see everyone moving away from him. It was, in the words of Nicholas of Cusa that I quoted in Chapter 1, a universe that had "its center everywhere and its circumference nowhere."

The notion of an expanding universe has become so much of a commonplace today that it is easy to forget how revolutionary it really is. In order to make this point clear, let us take up one by one the things about Hubble's view of the universe that are important philosophically.

THE UNIVERSE HAD A BEGINNING.

By regarding the expanding universe as a motion picture, you can easily imagine "running the film backward." If you do so, you find the universe getting smaller and smaller, and eventually you come to the moment when its whole mass is crammed into an infinitely dense point. From that point and that time to the present, the universe has been expanding. Before that time it didn't exist, or at least it didn't exist in its present form. The simple fact of universal expansion compels us to conclude that the universe had a beginning in time. We shall see that there is some controversy about its exact age, but for the present we simply note that most cosmologists would say that the universe has existed for about ten to twenty billion years. For scale, this can be compared to the age of the solar system (roughly four and a half billion years), the time since the disappearance of the dinosaurs (sixty-five million years), and the age of the human race (about three million years).

The event that marked the beginning of the universe was christened the Big Bang, the term that has now entered the vocabulary and common talk of our culture. Originally, the name referred only to the single initiating event; now, astronomers have come to use it to mean the entire picture of the universe decribed above—a universe that arose from a very dense initial state and has been expanding ever since. I shall be using the term "Big Bang" in this sense. It will refer to the whole process: initial event plus expansion. As for the initial event itself, let's call it the "moment of creation."

One common misconception about the Big Bang that should be disposed of right here is the notion that the universal expansion is analagous to the explosion of an artillery shell. The galaxies are not like bits of shrapnel speeding away from a central explosion. The raisin-in-dough analogy is a more satisfactory way to think about the whole process. What is expanding is space itself, not a cloud of galaxies within space.

The simple statement that the universe had a beginning in time is by now so obvious to astrophysicists that few of us give it a second thought. Yet it is a statement that has profound implications. Most civilizations embrace one of two opposite concepts of time, which we can characterize roughly as linear and cyclical, and even more roughly as Western and Eastern. Linear time has a beginning, a duration, and an end; cyclical time, as its name suggests, goes around and around forever.

In a universe that works on cyclical time, the question of creation never arises; the universe always was and always will be. The minute you switch to linear time you immediately confront the vexing question not only of the creation, but of the

Creator. Although there is no logical reason for the assumption, many people believe that if something comes into existence, it must do so in response to the actions of some rational being. Because of that belief, astronomers, even though they hate to get involved in theological discussion, find themselves in one when they posit the Big Bang universe. It puts them squarely in the middle of the age-old debate about the existence of God.

THE UNIVERSE IS NOT STATIC.

Until Hubble's discovery of the redshift, it was usually assumed that the universe was static, that the stars were eternal and unchanging, and that the problem of understanding the universe was akin to that of mapping out a newly discovered territory. The territory was intricate and the job difficult, but at least one could be sure that the land wouldn't move around while the map was being made. Hubble's discovery changed all that. Instead of a static universe we are dealing with a universe that is continually evolving in time. We must think of the universe as *process*.

Accordingly, all the observations made throughout recorded time give us little more than a snapshot of the universe during a small portion of its history. If the entire lifetime of the universe so far were compressed into a single year, recorded history would occupy roughly the last second before midnight on New Year's Eve. If the universe were static, then a snapshot would be enough to tell us how it has looked and behaved since the beginning of time. But since the universe is evolving, the snapshot tells us only how it is now, and we have to deduce how it has been in the past.

This point of view makes a great deal of difference in what we are willing to accept as explanations of what is found out there. We shall see later that one feature of the universe seems to be that it is full of large voids — regions in which no galaxies are found. If one takes the universe to be static and unchanging, then one has to find some mechanism by which such bubbles can be created and maintained through all eternity — an almost impossible task. On the other hand, if one believes the universe to be evolving, the task becomes much easier. All one has to do is think of something that can produce a bubble that will last long enough for us to see it. Many physical processes can produce a transitory bubble, as we shall see in Chapter 5.

THE UNIVERSE WILL NOT REMAIN IN
ITS PRESENT STATE FOREVER.

Just as the universe had a beginning, so it will have an end. We will go later into details about what form this end might take; for the moment we note that an evolving universe must necessarily evolve toward some end state.

In the nineteenth century, the notion of the "heat death" of the universe was in vogue. In this picture the universe would run down according to the second law of thermodynamics; the entire creation would eventually be reduced to a featureless, uniform blob of matter. The possible futures in the Big Bang cosmology are much more dramatic than heat death. That we cannot say with certainty which of the possible ends will occur is due to shortcomings in our observational techniques, and is probably just a temporary state of ignorance. But even now we can see that there are only two possible futures, and we can work out the details of each.

WHEN THE UNIVERSE WAS YOUNGER, IT WAS
DENSER AND HOTTER THAN IT IS TODAY.

When the universe was younger, the same amount of matter was packed into a much smaller volume than it occupies today. Consequently, its density was higher then than it is now. It is well known that when materials are compressed to a high density, they become hotter. The common everyday illustration of this fact is the use of a hand pump to inflate a tire. After a while, the pump barrel becomes hot, the heat being generated by the continued compression of the air. Similarly, when we go farther back in time we find the matter more densely packed and the temperature correspondingly higher. In fact, if I had to pick a single slogan I would want people to remember about the Big Bang it would be this: *As far as the universe is concerned, younger means hotter.*

This fact has enormous implications; indeed, it can be said to be responsible for the flowering of cosmology today. The reason is that when we say that a material is hot, what we mean is that its constituent atoms are moving quickly. The hotter the material is, the faster the atoms move. The faster they move, the more violent are the collisions that occur from time to time.

The situation is analogous to automobile accidents. In a cold material, the atoms move slowly and the collisions are like those of two cars bumping at low speeds in a parking lot. It's unlikely that the cars or atoms will be severely damaged. If a material is hot, on the other hand, the collisions will resemble cars colliding head-on at full speed. In that case, the cars and atoms are very likely to be torn apart, littering the landscape with fenders, bumpers, and other parts.

Thus, when the universe was hotter and younger the collisions between atoms were violent, and there must have been a time when the temperature was so high that no atoms could survive the collisions. All would be reduced to their constituent parts. From this we know that there must have been a time when no atoms existed and a time when they came into being. Before the creation of atoms, matter must have existed in the form of electrons wandering around looking for nuclei to which they could become attached, and nuclei wandering around in search of electrons — a state of matter that physicists call a "plasma." If an electron happened to hook up to a nucleus to form an atom, the two would be torn apart in the next collision.

This chain of events, in which matter proceeds from one state to another as the temperature falls, is something I like to call a "freezing," analogous to the transition of water from a liquid to a solid when the temperature drops below 32 degrees Fahrenheit. The transition from the mixture of electrons and nuclei to atoms may have taken place at a higher temperature than this, but the two processes have many properties in common.

WHEN THE UNIVERSE WAS YOUNGER, IT WAS SIMPLER.

An atom is a pretty complex structure. It has a compact, positively charged nucleus and a swarm of negatively charged electrons in orbits. A mixture of unattached nuclei and electrons, on the other hand, is a relatively simple system. After all, to make an atom you have to place all the constituent pieces together in just the right pattern. To make a mixture, you can just throw them together any which way. It's the difference between packing a suitcase carefully and just tossing everything in.

The story of what happens to atoms is typical of the early history of the universe. As the temperature keeps dropping as a result of the Hubble expansion, more and more complex structures are formed. The atoms, being the largest and airiest of the structures we want to consider, were the last to form. Moving backward in time, the next structure to go through a freezing were the nuclei themselves. Nuclei are simply aggregations of protons and neutrons. If they collide violently enough, the protons and neutrons can be knocked apart. Thus there must have been a time when the nuclei of atoms didn't exist and a time when they were born.

In the same way, the protons and neutrons and other elementary particles that make up the nuclei are thought to be made up of particles, called quarks, that are more elementary still.* When the temperature of the universe was very high, these quarks couldn't remain locked together in the elementary particles, but would come out as free particles. In other words, in earliest times there were none of the elementary particles that now make their home inside the nucleus. There was a time when they did not exist, and a time when they were born.

When the temperature was high enough, then, matter was a mixture of quarks and particles like electrons — particles that physicists call leptons ("weakly interacting ones"). According to our present ideas, this is the end of the line: matter can't be broken down any further. Everything around us is made up of different combinations of quarks and leptons. As the universe expanded, the quarks froze into the elementary particles, then the particles froze into nuclei, and finally the nuclei and electrons froze to make atoms. You must admit that it's a neat picture of the evolution of the forms of matter.

But the simplification of the universe doesn't stop with matter.

---

* The story of the elementary particles and the quarks is told in my book *From Atoms to Quarks* (Scribners, 1980).

Once matter has been broken down to its basic elements, there is still another source of complexity in the universe, and that is the fundamental forces which govern the way particles interact with each other. In our present, rather frigid universe, there are four such forces. These are (in decreasing order of strength) the strong force, which holds the nucleus together; the familiar force of electricity and magnetism; the weak force, which governs some kinds of radioactive decay; and gravity. Although these forces seem to be very different, according to our theories of the basic structure of the universe they are but different aspects of the same force. As the energy of collisions gets higher, the distinctions among the fundamental forces are supposed to disappear.

When the distinction between two forces disappears, we say that the forces become unified, and the theories that describe this process of coming together are called unified field theories. These theories have implications about the evolution of the universe. As we move backward in time beyond the point when matter was broken down into its ultimate constituents, the first forces to unify are electromagnetism and the weak force. Before this unification there were only three fundamental forces: the strong, the newly unified electroweak, and gravitational. Moving backward beyond this unification, the next significant event is the unification of the strong and electroweak forces. Before this unification there were only two fundamental forces: the strong-electroweak and gravity. Afterward, there were three. As this unification proceeds, the two fundamental constituents of matter, the quarks and the leptons, become interchangeable; in effect they become one sort of particle.

Finally, at temperatures so high that they are literally unimaginable, the theories suggest that the final two forces unify. (The theories that describe this process, the so-called theories of supersymmetry, will be discussed in Chapter 11.) For the moment, we need only note that they tell us that in its first fraction of a second of existence the universe was as simple and beautiful and elegant as it could possibly be. It consisted of a sea of one kind of particle, and these particles interacted with each other through only one kind of force.

As I enjoy telling audiences who have followed me thus far

in the history of the universe, it's all been downhill ever since then.*

The Timetable

The history of the universe since the moment of creation, then, has been one of expansion, cooling, and a gradual buildup of complexity as the temperature fell. I have not yet assigned any times to the events that mark the transition points — the "freezings" — in which either matter itself or the fundamental interactions change form. Without going into details, the development of the universe can be chronicled with the aid of a simple timetable. Let's assign to the initial event itself the quantity we call time zero. Then the following times mark the various freezings described above:

10^{-43} SECOND†

At this time the gravitational force separates from the strong-electroweak. After the short honeymoon of ultimate simplicity, the universe has become somewhat more complex, for interactions between particles are now governed by two kinds of forces. The temperatures are so high and collisions so energetic that they far surpass anything in the present universe, even the centers of stars and quasars. Some theorists have ideas about how things were at this time, but no experimental tests are available to see if they are right.

10^{-35} SECOND

The strong and electroweak forces separate, leaving the universe with three fundamental interactions. The quarks and lep-

---

* You can find a more complete discussion of the steps that are outlined here in my book *The Moment of Creation* (Scribners, 1983).
† I will be using the so-called scientific notation to save space. The notation is not hard to understand. A 10 with a negative number as an exponent should be interpreted as a command to "move the decimal point this many places to the left." Thus, the number above is a decimal point, forty-two zeroes, and a 1. A positive exponent means "move the decimal point this many places to the right."

tons stop being interchangeable and assume pretty much their present form. In addition, in this freezing a number of strange objects may have been formed, objects whose creation required much more energy than is now available either in nature or in the laboratory. These objects, such as the cosmic strings discussed in Chapter 12, are stable once they are formed. They could therefore have survived since this very early time and may still play a major role in the structure of the universe.

On the experimental front, the theories that describe this transition do not show a strong record. Called the GUT ("Grand Unified Theories"), they predict that the proton, hitherto thought to be completely stable, will decay on a time scale much longer than the lifetime of the universe. Attempts to see this decay with extremely sensitive experiments have so far come up dry, so that the lifetime must be much longer than that predicted by the simplest versions of the theories.

10^{-10} SECOND

This marks the last freezing of the forces. The electromagnetic force separates from the weak, and the universe is left with its full complement of four fundamental interactions. This also represents the earliest time for which we can recreate the conditions of the universe in our laboratories. In giant particle accelerators, the conditions during this epoch can be produced over volumes the size of a proton. It doesn't sound like much, but it's enough to give us a very firm handle on what happened.

10 MICROSECONDS

The quarks coalesce to form the elementary particles.

THREE MINUTES

Protons and neutrons come together to form nuclei. Actually, only light nuclei — up to helium and lithium — are formed during these early stages of the universe. All heavier elements are made later, in stars.

100,000 TO 1,000,000 YEARS

Electrons and nuclei come together to form atoms. Once the atoms formed, the universe had achieved something like its present form, and the Hubble expansion has gone on without any more radical changes.

Evidence for the Big Bang

The Big Bang picture, as anybody can see, is a pretty comprehensive view of the evolution of the universe, taking us from the moment of creation to the present in a series of relatively simple steps. But is it really what happened, or is it just a plausible tale, like something out of Rudyard Kipling's *Just So Stories*? The only way to answer this question is to look at the observational evidence that supports the theory. Other than the redshift itself, there are two main pieces of evidence available: Big Bang nucleosynthesis and the cosmic microwave background. There are a number of other pieces of evidence, but they are more difficult to describe and, on the whole, less impressive than these two.

Nucleosynthesis (the putting together of the nuclei) refers to the processes which occurred at the three-minute milestone. For a brief period, protons and neutrons collided with each other and stuck together, forming the light nuclei. Before three minutes, the temperatures were too high to allow nuclei to stick together, and afterward the Hubble expansion had proceeded to the point where the density of particles was too low for many collisions to take place. Thus there is a very narrow window around the three-minute mark during which enough collisions take place to form a significant number of nuclei and the temperature is low enough to allow the newly formed nuclei to survive.

The theory of the Big Bang tells us how densely packed matter was during this brief period, and hence how many collisions there must have been. The collisions themselves can be reproduced in our laboratories, so we know how often each collision will produce a given nucleus. Thus the number of each type of nucleus produced — the "primordial abundance" — becomes a critical test of the whole Big Bang picture.

The best example of how this test works relates to the primordial abundance of helium-4, a nucleus with two protons and two neutrons. The theory predicts that 25 percent of the matter in the universe after the three-minute mark should consist of this element. When astronomers look out and measure the actual amount of helium in the universe today and then subtract the amount that has been made in stars since the Big Bang, they arrive at almost exactly the predicted amount. Had they found an abundance that differed by as much as 2 to 3 percent from the prediction, the Big Bang theory would have been in serious trouble.

This story of precise prediction and subsequent verification can be repeated for a number of different nuclei, including deuterium (one proton, one neutron), helium-3 (two protons, one neutron), and lithium-7 (three protons, four neutrons). They all agree; the predictions of the Big Bang are borne out whenever they are tested.

The second dramatic bit of evidence for the Big Bang comes from a completely different source. It is best understood in terms of an analogy. If you come into a room after a fire has gone out, you can usually tell how long it is since the fire was burning by looking at the coals. If they are red-hot, the fire must have gone out recently. If they're dull orange, it probably went out a half hour ago. If they're gray (i.e., if they're not giving off visible light) but you can still feel radiation with your hand, then the fire is older still. In this progression, the cooling coals give off radiation of progressivly longer and longer wavelength, from visible light to infrared (which you can't see, but can detect with your hand).

The early stages of the Big Bang can be thought of as a fire and the universe itself as the coals of that fire. Like the coals in your fireplace, the universe has been giving off radiation of increasingly longer wavelength as it has cooled. Today, some fifteen billion years after the fire last burned, that radiation must be in the form of microwaves — the same kind of radiation we use to cook and to send TV signals. The only difference between the universe and the coals in the fireplace is that in the latter case we stand outside the coals and sense the radiation, while in the former we are inside the body that is doing the radiating; in effect, we are inside the pile of coals itself.

In 1964, Arno Penzias and Robert Wilson, two physicists working at Bell Telephone Laboratories in New Jersey, turned a large radio receiver toward the sky and discovered that no matter which way they pointed it, they were getting a signal indicating the presence of microwaves. After some discussion in the scientific community, it was realized that this radiation was precisely what one would expect if the universe had started as a hot Big Bang roughly fifteen billion years ago. In that time, it has cooled from the high temperatures of the beginning to the temperature characteristic of microwaves — about 3 degrees above absolute zero. It was probably this discovery, coming as it did at a time when the structure of the universe was still very much an open subject, that tipped the balance and swung the scientific community over to its present support of the Big Bang.

What Are the Odds?

In the winter and spring of 1986, I had the privilege of spending a sabbatical leave working with the paleontology group at the University of Chicago. I was working on the problem of mass extinctions ("What killed the dinosaurs?"), and the visit gave me a chance to get acquainted with David Raup, the leader of the group. Dave has been described as the most brilliant paleontologist in the world, an evaluation that, after a period as his collaborator, I think is well deserved. He is also one of those rare individuals who get tremendous enjoyment from everything they do. He loves to question things that everyone else accepts without question, a trait that I suspect has a lot to do with his success as a scientist. He's also a terrific cardplayer (as I learned to my sorrow at some late-night poker sessions), and spends some of his spare time at his personal computer trying to find a way to beat the system at blackjack.

I mention these things as background to what I want to tell you about — something that happened during my visit with the group. We were in Dave's office discussing something or other when he turned to me and asked, right out of the blue, "What are the odds that the Big Bang is correct?"

That brought me up short, as you can imagine. My first impulse was to say, "Of course it's correct," but I suspected he'd ask me how I knew, so I paused. The more I thought about the question, the more fragments of memory floated into my mind. I recalled a bag lunch in a prestigious physics department when a prominent senior member of the faculty (I wouldn't dream of embarrassing him by mentioning his name) said that he often thought of leaving a sealed envelope to be opened fifty years after his death. In the envelope would be predictions about the way certain scientific controversies would turn out. At the top of the list would be the prediction that the interpretation of the redshift as evidence for a universal expansion would turn out to be wrong.

This memory, together with my experience of fads and fashions in cosmology that we'll discuss later, made me reluctant to be too dogmatic in my reply. Finally, I said that I thought the general picture of the universal expansion from a hot beginning had a very high probability of being right, but that many of the

ASPECT OF THE THEORY	PROBABILITY
General picture of universal expansion from a hot beginning	99% plus
Formation of atoms and nucleosynthesis	95%
Electroweak unification*	95%
Quark freezing and the general quark-lepton picture	85%
Strong-electroweak unification	50%
Unification with gravity	30%
Supersymmetry, superstrings	20%
Any of the theories of galaxy formation and large-scale structure discussed later on	10%
*This must be right—they've already given out the Nobel Prizes for it.	

details of our understanding could easily be wrong. This answer seemed to satisfy Dave, who apparently had thrown the question at me because he had asked the same thing of a member of the Chicago astronomy department and been told that the Big Bang was 100 percent certain. Like any skillful gambler, Dave distrusted a sure thing.

But the more I thought about the question, the more I realized that it is not possible to give a single, definitive answer to it. The Big Bang theory has many facets, and each has a different level of confidence attached to it. As an exercise in making judgments, in the chart opposite I give my own guess at the odds that various aspects of the Big Bang picture presented in this chapter are correct.

FOUR

Five Reasons
Why Galaxies Can't Exist

The progress report is that there's no progress.

—NICK KOSOROK,
Montana snowplow driver,
during the blizzard of '85

W E CAN SUMMARIZE the modern view of the universe in two brief statements. First, the universe has been expanding ever since it was formed, and in the process has evolved from simple to complex structures. Second, the visible matter in the universe is organized hierarchically: the stars grouped into galaxies, galaxies into clusters, and clusters into superclusters. The problem we face, then, is to understand how a universe whose evolution is dominated by the first statement could become one whose structure is described by the second.

The problem of explaining the existence of galaxies has proved to be one of the thorniest in cosmology. By all rights, they just shouldn't be there, yet there they sit. It's hard to convey the depth of the frustration that this simple fact induces among scientists. Time after time, new developments have come along and it has seemed that the problem was solved. Each time the solution turned soft, new problems developed, and we were right back where we started.

Every few years the American Physical Society, the professional association of physicists, has a session at one of its meetings where astrophysicists talk about the hottest new ways of tackling the galaxy problem. I have been a bemused listener at many of these sessions, and I have come away with a great respect for the ingenuity of my colleagues. At the same time, I have also come away with a profound skepticism about the ideas they advance. Having sat through explanations of how turbulence, black holes, explosive events during galaxy formation, heavy neutrinos, and cold dark matter would solve all our problems, I've developed an immunity to what Princeton astrophysicist Jim Peebles calls the "snake-oil approach" to cosmology. Despite what you may read in the press, we still have no answer to the question of why the sky is full of galaxies, although we've suceeded in eliminating many wrong answers. We may be a lot closer to the truth now than we were before, but don't let that lull you into gullibility. It is not impossible that the details of the solutions described later on will join the failed efforts of the past. We shall argue, though, that some of the general ideas implicit in these solutions will very likely form part of the grand solution — if we ever find it.

To start with, I want to impress this persistent difficulty upon you by giving five reasons that galaxies cannot exist. This will also give us a chance to stroll through the Rogue's Gallery of failed ideas — those ideas that didn't work. This should leave us with an acute sense of humility as we examine the theories of our own time.

REASON #1: GALAXIES COULD NOT HAVE FORMED BEFORE ATOMS.

We may think of the universe during the early stages of the Hubble expansion as made up of two constituents: matter and radiation. We saw earlier how the matter undergoes a series of freezings as it gradually builds up to more and more complex structures. As these changes in the form of matter proceed, the way in which radiation and matter interact changes radically. This, in turn, plays a pivotal role in the formation of galaxies.

Light and other kinds of radiation interact strongly with free, electrically charged particles of the type that existed in the plasma that made up the universe before atoms formed. Because of this interaction, when radiation moves through such a plasma it collides with particles, bouncing back and exerting a pressure in much the same way that air molecules bouncing off the wall of a tire keep the tire inflated. If it happened that a galaxy-sized conglomeration of matter tried to form before the freezing of the atoms, radiation streaming through the material would have blown the conglomeration apart. By the same token, the radiation would tend to be trapped inside the matter. If it tried to get out, it would suffer collisions and bounce back.

The importance of this statement is hard to overestimate. What it means is that so long as matter remained a plasma (i.e., so long as atoms had not frozen out), no galaxies could have formed or even begun to form. It follows that there was a definite period, starting around 100,000 years, when the onset of galaxy formation occurred. Before that time, the interaction of radiation with matter would have prevented anything like our present universe from forming.

After atoms had formed, the situation would have been markedly different. The key fact here is that radiation does not interact as strongly with atoms as it does with the charged particles in a plasma. You can consult your own memory for evidence that this statement is true. If you have stood on a mountain peak or a high building and looked out over the surrounding landscape, for example, you have probably been able to see landmarks fifty or even one hundred miles away. In some places, such as the mountaintops rising in the clear air of the western United States, you can see even farther.

Now before you see these landmarks, it is necessary that light should travel from the object being seen to your eye. The simple experience of seeing a long way, then, tells us that light can travel long distances through the air without being scattered or disturbed. This cannot happen in a plasma. That it happens in air, which is made up of atoms and molecules, shows that the interaction of light with these two forms of matter must be very different.

In the early universe, then, the sequence of events must have gone something like this. Up to about 100,000 years, matter was in the form of plasma, and no galaxy-sized objects could have formed. At 100,000 years, the first atoms started to appear and the interaction of light with matter began to weaken. The formation of the atoms did not take place all at once, but continued up to the million-year mark. Between these two times, the makeup of the universe shifted gradually from plasma to atoms, and by the time the transition was over, few free charged particles were left. The dominant form of matter was the atom.

Sometime during the formation of atoms, the strength of the interaction between matter and radiation dropped to the point where the latter was no longer trapped inside the plasma. The radiation streamed out freely, and from that point on had little effect on the process of galaxy formation. In the jargon of cosmologists, we say that during the formation of atoms, radiation *decoupled* from matter.

Although the decoupling was gradual, I shall occasionally want to make a loose reference to the process. I shall speak of it as occurring at around the 500,000-year mark, since this is a round number roughly halfway through the freezing of the atoms. This will be for shorthand use only; it should not be taken to mean that the universe was opaque up to 500,000 years and turned transparent at 500,000 years plus one second.

I have found a useful analogy for helping to visualize the process of decoupling. When you have a drink like iced tea which comes in a tall glass, watch what happens when you stir in the sugar. At first, the drink becomes cloudy, because at that point the sugar is in the form of relatively large lumps, and large lumps scatter light efficiently. You know that the scattering is efficient because light cannot get all the way through the glass, but is scattered instead. It is this scattered light that gives the tea its cloudy appearance. In this state the tea is analogous to the universe before the formation of atoms, when radiation was interacting with the plasma. After a few moments, the tea suddenly becomes transparent again. What has happened is that the sugar has dissolved and now exists in the form of molecules which interact weakly with light. The light now comes through the tea without being scattered, and the cloudiness is gone. This change from cloudiness to transparency in your tea resembles what happened in the universe when atoms formed. The universe became transparent as the radiation decoupled, and there was nothing left to counteract the force of gravity pulling the matter together.

So the interaction of radiation and matter prevents the beginning of processes that could lead to galaxies before the universe is 500,000 years old. This turns out to be a major problem, because of . . .

REASON #2: GALAXIES HAVEN'T HAD TIME TO FORM.

Gravity is the great destabilizing force in the universe. It never lets well enough alone; it's always acting, trying to pull clumps of matter together. In a sense, the entire history of the universe can be thought of as a desperate and ultimately futile attempt to overcome gravity. It would be amazing, given the universal nature of the gravitational force, if it had not played a major role in the formation of galaxies.

Suppose the universe had begun as a uniformly smeared-out collection of matter, no place having a greater concentration of mass than any other. In this situation, you could expect that the force of gravity might act to pull everything in the universe together into one impossible central sun. You might think this, but you'd be wrong.

The problem is that in any collection of matter, however smoothly it is distributed, there will be slight concentrations somewhere. Even if we have to go down to the microscopic level to find it, the random motion of the atoms will eventually result in a state of affairs where there is a small excess of atoms in some spots and a small deficit in others.

It is not hard to visualize what happens next. At a given moment, a little extra matter accumulates somewhere, either because of atomic motion or for some other reason. Because of the momentary excess of matter at that point, the gravitational force exerted is greater than that exerted by surrounding points. Consequently, more mass will be drawn into the area where the original concentration occurred. With more mass, the concentration is able to exert still more gravitational force and attract still more matter to it. No matter how smooth the initial distribution is, once the smallest concentration forms, the uniform blob will break up into smaller pieces, each built around one of the original mass concentrations. This inherent instability of a mass of matter was first pointed out in the 1920s by the British astrophysicist Sir James Jeans.*

At first glance, this seems like a ray of hope. The universe *must* break up into small units of mass, and, with luck, those units could turn out to be galaxies. It even turns out that although we have talked only about a nonexpanding universe, the Jeans result is just as true if there is a Hubble expansion. But the problem isn't that simple. The same theory that tells us that a smooth distribution of matter is unstable against breakup into small chunks also tells us how long the breakup process will take.

It comes down to this: Can the gravitational forces act quickly enough after decoupling occurs to gather matter into galaxy-sized clumps before the Hubble expansion carries everything out of range? One of the great shocks to the astronomical community in the 1930s was that the answer to this question is a resounding "No!" What seemed the likeliest mechanism for galaxy formation—the mechanism of gravitational instability we have just described—will not work in an expanding universe. Perhaps it was this fact that led Jeans, later in life, to propose a universe in which matter was being created continuously in the voids left behind by galactic expansion. In this picture, galaxy formation is a continuous process, not confined to any particular time in the history of the universe. Later codified as the Steady-State Universe, Jeans's theory was eventually

---

* What Jeans actually showed was that a gravitating mass is unstable against breakup into chunks of a certain size. If a mass is smaller than the smallest chunk into which it could be divided, the mass will be stable. Otherwise, it will break up. This test is known as the Jeans criterion.

For experts, the Jeans length is $\left\{ (\pi/G\rho)\dfrac{dp}{d\rho} \right\}^{1/2}$.

abandoned after convincing evidence in favor of the Big Bang had been accumulated (see Chapter 3).

So the problem of galaxy formation can be stated as follows: Galaxies cannot begin to form until after radiation and matter decouple. If, however, the only mechanism at our disposal is gravitational instabilities of the Jeans type, all the matter will be carried out of range before anything like the present galactic masses can collect. There is a narrow window in time between decoupling and the point when matter is too thinly spread, and any galaxy-formation mechanism we can accept has to work quickly enough to fit into this window.

One way out suggests itself. If we didn't have to wait for the atomic concentrations to build up, if we could find some way of giving the gravitational collapse a running start, we might be able to squeeze galaxy formation through in the time allotted. One possible way for this to happen is to have the mass concentrations formed by some other physical process, such as turbulence in the gas clouds after the formation of atoms. Alas, this line of argument leads us to . . .

REASON #3: TURBULENCE WON'T WORK, EITHER.

The "running start through turbulence" is a simple idea, the first versions of which were aired around 1950. The postulate is this: Any process as violent and chaotic as the early stages of the Big Bang is not going to result in a smooth distribution of matter. The Big Bang will not be like a deep, placid river, but like a mountain stream, full of whitewater and turbulence. In this chaotic flow we can expect to find eddies, swirling vortices of gas. In this theory, an eddy is, in effect, a mass concentration of the Jeans type, pulling in surrounding matter because of its gravitational attraction. If the eddy is of the right size, it can pull together a galaxy-sized mass before it has a chance to dissipate. By then that mass will be big enough so that it will be held together by the force of gravity after the eddy is gone.

Neat, but there are some difficulties. In the first place, an eddy that forms before the 500,000-year mark is still a mass concentration, and like every other mass concentration it will be blown apart by radiation pressure. Consequently, the turbulent eddies that serve as condensation nuclei for the galaxies must come into existence after the appearance of atoms. What this means is that the eddies that form right after the atomic freezing are the ones most likely to lead to galaxies, because they are the ones that have the most time to gather matter together. If these eddies are of the right size, they could indeed produce the galaxies we see. We could just assume that there were eddies of galaxy size (or near galaxy size) present at the time of the freezeout.

But this approach raises an awkward philosophical point. We can look at the galaxies we see, extrapolate backward in time, and posit a set of turbulent eddies that will produce them. This doesn't solve the problem, it only puts the old question differently, pushing it back one notch. Instead of asking the question "Why are galaxies as they are?" we ask the question "Why were the eddies as they were?" Not much progress, is it?

In any case, the idea of using turbulence to get the galaxies started just didn't work. The lifetime of the eddies — the duration of their swirling existence — isn't long enough to produce the kinds of galaxies we see. This approach to galaxy formation was dropped by the middle 1970s.

REASON #4: GALAXIES HAVEN'T HAD TIME TO FORM CLUSTERS.

Perhaps we are encountering difficulties because we are looking at the galaxy problem in too narrow a focus. Perhaps what we should do is look at things on a larger scale and hope that if we understand how clusters of galaxies form, the genesis of individual galaxies will take care of itself. This idea leads us naturally to the question of how very large mass concentrations could have formed early in the lifetime of the universe. One of the simplest ideas about what the universe might have been like when atoms were forming is that whatever else was going on, the temperature was the same everywhere. This is called the isothermal model. It corresponds to assuming that the radiation in the early universe was spread out smoothly, whether matter was clumped together or not.*

If you work out the mathematical consequences of the isothermal model, you find that the kinds of mass concentrations that would have formed in the early universe are very easy to describe. With the same temperature everywhere, ordinary random fluctuations would produce mass concentrations of all sizes; if you wanted to find a planet-sized concentration, there it would be. The same holds for star-sized concentrations, galaxy-sized, cluster-sized, and so on. In the jargon of the astrophysicist, mass concentrations would appear on all scales.

In this model there is a particularly simple solution to the galaxy problem, because the smallest mass concentrations grow faster than the larger ones. The first objects to accrete would be relatively small things called protogalaxies, containing perhaps a million stars each. These protogalaxies would then clump together under the influence of gravity to form full-fledged galaxies, which would then clump together to form clusters and superclusters. The universe, in this model, builds itself "from the bottom up."

The only hitch is that there simply hasn't been time for the leisurely bunching under the influence of gravity to have taken place since the moment of creation. Yet, as we shall see in detail in Chapter 5, there are some very large and complex collections of galaxies in the sky. We are forced to conclude that the universe could not have had a constant temperature throughout when decoupling occurred.

The argument we've just presented isn't, strictly speaking, an argument against the existence of galaxies. It only shows that galaxies can't exist if we assume that radiation was smoothly distributed in the early universe. This assumption, while reasonable enough, is not written on tablets of stone. Since it doesn't work, we are always free to try something else — for example, the assumption that radiation was *not* distributed uniformly in

---

* The connection between constant temperature and smoothly distributed radiation may not be obvious, but it would take too long to prove it. Please take it as a given for the moment.

the early universe. We proceed to do so and thereby collide
with . . .

REASON #5: IF RADIATION CLUMPS
WITH MATTER, AND MATTER CLUMPS INTO
GALAXIES, THE COSMIC MICROWAVE RADIATION
COMES OUT WRONG.

If radiation was not spread out smoothly, independent of the
matter in the universe, where could it have been? Following the
standard procedure of theoretical physicists, we consider next
the opposite assumption. We assume that in the early universe,
matter and radiation were grouped together. If this was so,
whenever a mass concentration was found, there would also be
a concentration of radiation. In the jargon of physics, this is
called the "adiabatic" situation. It will arise whenever changes
in the distributions in the gas take place so fast that energy
cannot be transferred easily from one point to the next.

We know that in order to make galaxies, the matter in the
universe had to be pretty well segregated into clumps when the
atoms formed. We called this "giving the process a running
start." A necessary corollary is that under adiabatic conditions
the radiation must have started out being clumped up as well.

But this result flies in the face of one of the most remarkable
facts about the universe that we know. If you look at the microwave radiation (see Chapter 3) coming to us from the direction of the earth's north pole, and then turn around and look
at the radiation coming from the direction of the south pole,
you find that they are almost exactly identical. In fact, no matter
which way you look in the sky, the background radiation coming in is identical to what you see in any other direction. This
statement is true to an accuracy of one part in a thousand. From
this remarkable uniformity we conclude that when radiation
decoupled from matter it had to be fairly uniformly distributed
throughout the universe.

The upshot is this: What is required of the microwave background by the galaxy formation process and what we observe
of its uniformity are diametrically opposed to each other. The

former requires radiation to be bunched with matter, so if matter was clumped when atoms formed, there should be traces of that clumping in the cosmic microwave background today. On the other hand, the observed uniformity of the microwave background implies that radiation could never have been so thoroughly clumped; if it had been, it wouldn't be uniform today. When detailed numerical calculations are done, astrophysicists find that it is impossible to reconcile these two conflicting requirements. The microwave radiation can't be uniform and nonuniform at the same time.

What Now?

The reasoning above demonstrates very clearly that we cannot take a universe full of galaxies for granted. Explaining that universe has turned out to be much more difficult than anyone would have guessed back in Hubble's time. But our examination of failed theories has also shown some of the elements that a correct theory of galaxy formation must contain.

We know that if galaxies have to wait until after radiation decouples before they can start forming, they'll never make it. Gravitational collapse from a uniform distribution of matter is too slow to counteract the Hubble expansion. It follows that the universe has to come out of the decoupling with the galaxies already well along on the way to completion. They need not be preformed, but at the very least the universe must be seeded with some kind of mass concentrations that can trigger the process of gravitational collapse. These concentrations would act like the bits of dust that serve as the nuclei around which raindrops form in the atmosphere — they would be the condensation nuclei for the galaxies.

If these nuclei are to be there when needed, they have to be made at some point early in the history of the Big Bang and survive until the 500,000-year mark. From the discussion on page 56 (Reason #1), we know that ordinary concentrations of matter could never do this. They would be blown apart by radiation pressure long before they could serve as a condensation nucleus for a galaxy. Whatever "seed" was made early on must be able to survive the buffeting of the radiation for long periods. The "seeds" must, therefore, be made of some type of matter that does not interact strongly with radiation. It is this hint about the universe, as we shall see in Chapter 7, that provides us with some hope that we will be able to find our way out of the kinds of dilemmas we have been discussing.

Before turning to that question, however, we should note that reasons #4 and #5 suggest strongly that the galaxy problem is in reality a small part of a much larger question—the question of the overall structure of the universe. Before moving on to consider modern solutions to the problem, then, we will take some time to discuss in more detail the way that galaxies are arranged in the sky.

FIVE

Bubbles and Superclusters

And I know that my life's been a failure
Watching the bubbles in my beer.

—Country-and-western song

THE ART INSTITUTE OF CHICAGO has one of the world's best collections of late nineteenth-century French painting, collected during the time when the city really was "hog butcher to the world, player with a nation's railways." One of the more popular works is a large canvas by George Seurat. Its formal title is *La Grande Jatte*, but it's known colloquially as *Sunday Afternoon in the Park*. It shows a group of Parisians strolling around a park next to the Seine. The painting was even the inspiration for a Broadway show called *Sunday in the Park with George*. Seurat used a painting technique that was rather unusual in his time. Instead of drawing his brush over the canvas in the customary way, he touched the canvas only with the tip. The result is a painting made up of a large number of small dots of color. This style of painting is called pointillism.

67

Because of this technique, looking at the painting is an unusual experience. From far away, you see what the painter intended —the scene of a park with figures in it. If you go up close, however, the scene disappears and all you see is a collection of colored dots on canvas. The smoothness that is apparent when you look at the "big picture" hides the actual dotted appearance of the composition.

Seurat's painting supplies a useful analogy for one of the most cherished notions astronomers hold about the structure of the universe—the notion that if we look at a large enough scale, we will find the universe to be smooth and homogeneous. From Einstein on, leading cosmologists have assumed this statement to be true.

Yet the fact of the matter is that the universe we know is lumpy. In our own neighborhood, it consists largely of empty space, interrupted only by the sun, the planets, and the bits of rock we call asteroids. Looking outward, we find a universe in which the visible mass is clumped into galaxies separated by large distances, the galaxies themselves being gathered into clusters. Try as we will, there seems to be no way to get a large enough view of creation to see a simple, smooth structure. We just can't seem to get far enough away from the painting.

Scientists find this state of affairs irritating. I have always been dimly aware of my colleagues' preference for homogeneity, but until I started planning this book I never gave much thought to the reasons why they feel this way. I don't think it is analogous to the assumption that made the Greeks prefer geocentrism. Scientists can certainly see inhomogeneity if it's there and deal with it if they have to. It's just that they hope, in their heart of hearts, that at some level it will disappear from the universe as Seurat's dots disappear if you get far enough away from his painting.

Part of the reason for this tendency is historical. Physics and astronomy developed by studying homogeneous systems, and there's no question that they're the easiest thing to handle. Even when scientists deal with systems made up of discrete units like galaxies or atoms, they prefer to ignore the graininess and pretend that the structure is smooth. This way of dealing with nature has one enormous advantage: It works. If it didn't, engineers would never have been able to build a steam engine or pipeline before the atomic theory was developed. We can usually ignore the fact that water is made of atoms and treat it as if it were a continuum. This fact is drilled into every science student from the very beginning of his or her education. It's no wonder, then, that we have developed a kind of feeling—nostalgic though it may be—for the days when everything was smooth, and that we keep hoping our universe will eventually be found to fit the familiar mold.

This prejudice survived the discovery of galaxies; it was always possible to believe that we were still too close to the picture. But since the late 1970s, it has become increasingly difficult to hang on to the faith. As we look at the universe at larger and larger scales, we fail to see a system becoming increasingly simple. In fact, we see a universe that seems to remain complex out to the largest scales we can make out. Two discoveries brought this point home forcefully to the scientific community: the discovery, beginning in 1978, of a series of very large galactic superclusters, and the discovery of voids in 1981.

Clusters and Superclusters

We have already alluded to one of the important structural features of the universe—the fact that galaxies are not randomly distributed in space, but are grouped together in what are called galactic clusters. The first serious study of clusters was done by the late George Abell of the California Institute of Technology in 1958. Working from photographic plates taken at Mount Palomar, he identified 2,712 groupings of galaxies now known as Abell clusters. These are structures in which large numbers of galaxies are found in close proximity to each other. Over half of all galaxies in the sky are associated with clusters of various sizes.

The Milky Way, for example, is part of what astronomers call the "local group." This particular cluster consists of the Milky Way and Andromeda, two massive galaxies that play the role of gravitational anchors, and at least twenty other smaller galaxies that are dragged along with them. The entire assemblage is about 3 million light-years across. As clusters go, this is not very impressive. Some of the larger ones have many thousands of galaxies in them.

The Discovery of Large-Scale Structure

The story to be told in this section can be summed up in a very simple way: Nearly all the luminous matter in the universe — the matter that we can see — is contained in superclusters of galaxies. These superclusters are like long strings on which clusters of galaxies are strung like pearls on a necklace. The spaces between these superclusters, called voids, are relatively free of luminous matter, as we shall see presently.

The idea that superclusters might exist started with the work of George Abell described above. Although it wasn't a major field of research, the 1960s and early 1970s saw a steady stream of attempts to use something like Abell's technique to deduce the large-scale structure of the universe. In 1967, Donald Shane and Carl Wirtanen at the Lick Observatory in California published a catalogue of the locations of a million galaxies in the sky — by far the most comprehensive survey that had ever been done. Working entirely with the unaided eye, the two examined thousands of photographic plates and recorded the positions of the galaxies on them.

Suppose you were assigned the task of examining the street map of every large city in the United States and were told to write down the locations of all major street intersections. Your task would be comparable to the one Shane and Wirtanen accomplished. No wonder it took them twelve years to do it! According to the folklore of astronomy, they developed all sorts of ways of sneaking in their counting in the midst of other chores. Wirtanen, as lab director, is even supposed to have been able to work on a plate while carrying on the kind of interminable phone conversations that administrators must put up with in return for their fancy offices.

Using the Shane-Wirtanen survey, P. J. E. ("Jim") Peebles and his co-workers at Princeton put together a map of the universe. This was the first visualization of the universe in the large, and as such it has enjoyed a wide popularity. It appears in all sorts of textbooks and as a poster in almost every astronomy department in the country; it was even marketed as a design for the inside of a soup bowl. From our point of view, the striking thing about the map is that it seems to show in a very dramatic way the stringy, lacelike structure of the universe.

But looks can be deceiving. When Peebles's map was first published, there was a grand debate among astronomers about how to interpret it. Since the photographic plates from which it was derived are two-dimensional images of the heavens, there is no guarantee that the galaxies that seem to lie along a filament are really connected with each other. They could be merely lying along the same line of sight from the earth and be otherwise unrelated.

There is another fact to be noted about the "Million Galaxy Map," a particularly informative one which touches on a facet of the human brain that is little appreciated. Our vision and our brains are well adapted to seeing patterns in our environment. To our distant ancestors the primates, this sort of skill was essential for tasks like finding edible fruit in a tangle of leaves. We carry this heritage with us in the magnificent capacity of the human brain to detect patterns. No computer yet built can match a three-year-old when it comes to this sort of thing. We are in fact so good at seeing patterns that we often see them even when they're not there!

Think of the Rorschach test. Developed by psychiatrists to aid them in their studies of the human personality, it consists of a series of cards on which amorphous ink blots are shown. There is no intrinsic pattern in the blots, but when you look at them you see things. Each person's mind imposes its own patterns on the blots, which is what gives the psychiatrist an insight into the way that mind operates. Going back to the galaxy map, what are we to make of that lacework pattern? Is it really there, or has the universe presented us with a large-scale Rorschach test?

This is not an abstract philosophical question — it can actually be answered by applying some fairly simple statistics. The scheme is sketched in Fig. 5.1 (left) on page 72. Center your attention on a single galaxy, such as the one labeled A. Then count how

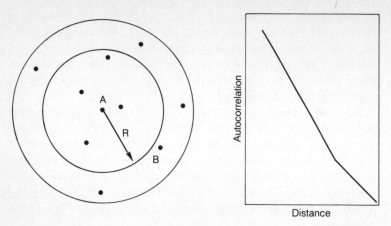

FIGURE 5.1

many other galaxies fall less than a distance R away from A. In the figure, there are four such galaxies. You can repeat this count for different values of the distance R, casting your net ever wider as you go. This operation is symbolized by the concentric circles in the figure. When you have finished this job for galaxy A, you transfer your attention to another galaxy, such as the one labeled B, and repeat the performance. You're not through until you've done this for every galaxy in the collection.

When the task is finished (needless to say, it's carried out by a computer), what you have is something that mathematicians call an autocorrelation function. It tells you what the probability is that two galaxies will lie within a given distance of each other. From this probability you interpret the galaxy map as follows:

If the probability is large for small values of R, this means that galaxies are likely to lie near each other, and this would be evidence for clustering. If no particular value of R had a probability significantly bigger than any other, the galaxies would be scattered more or less randomly throughout space — no clustering. Fig. 5.1 (right) sketches an autocorrelation function for the Shane-Wirtanen plot. It clearly shows the effects of clustering — it is largest for small values of R. Since this test is independent of human perceptions, we can safely rule out the alternative of a cosmic Rorschach test.

Before going on, I should point out that a major battle has broken out recently over another aspect of the Shane-Wirtanen survey. The main critic is Margaret Geller of the Harvard-Smithsonian Center for Astrophysics. Her argument is based on the fact that when Shane and Wirtanen did their survey, they broke the sky into squares and then parceled out the squares so that each man did every other one. Geller's point is that if one observer had different criteria from the other in examining the plates (for example, if one would accept a particular fuzzy speck as a galaxy and the other wouldn't), the result would be a pattern of alternate light and dark spots on the galaxy map. Such a pattern could easily masquerade as a filamentary structure.

Defenders of the plot, most notably P. J. E. Peebles, have countered with a series of complex and rather technical calculations, and the whole question seems to be sinking into a morass of statistics and mathematics of interest only to experts. I mention the conflict only to emphasize how difficult it is to provide a final answer to the question of the large-scale structure of the universe using the ordinary photographic techniques of traditional astronomy. This point should be borne in mind when, later in this chapter, we introduce the idea of the redshift survey. As we shall see, these surveys supply a much more reliable measure of the way that galaxies are grouped in the sky.

Several major redshift surveys done since the late 1970s have turned up over a dozen long, stringlike structures called superclusters. The largest of these stretches across the constellations of Perseus and Pegasus—constellations that are prominent in the evening sky during the fall and early winter. This supercluster is the largest known structure in the universe. In the drawing in Fig. 5.2 (page 74) we show a sketch of this supercluster based on a model made by David Batuski when he was at the University of Arizona. It consists of forty-three galactic clusters strung together on a couple of branching filaments, with the filaments themselves surrounded by the large empty spaces we call voids.

To me the most fascinating thing about Fig. 5.2 isn't the precise form of the supercluster, but the sheer size of it. It stretches a full billion light-years across the sky. This is an appreciable fraction of the total size of the universe. That we have already

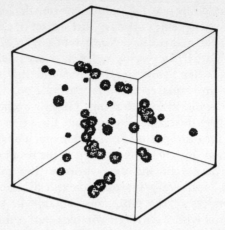

FIGURE 5.2

found structures of this magnitude seems to me to be a strong forecast that the universe is going to be lumpy and stringy all the way out to the limits of our observation. I can see little hope that homogeneity will suddenly set in at scales above a billion light-years.

Is the Perseus-Pegasus supercluster the largest structure in the universe? It is certainly the largest seen so far, but one must remember that all the surveys yet made have explored less than 1 percent of the volume of the universe within a billion light-years of the earth, and considerably less than 1 percent of the total observable volume. It would be amazing if in the infancy of our systematic surveys of the sky we stumbled at once across the biggest thing there was. Even beginner's luck must have its limits, and this means that we can expect to find even larger structures as our search broadens.

Voids

Voids are just what the name suggests: large regions of space in which either no galaxies or very few exist. Voids are huge —sometimes as much as 250 million light-years across. Under normal circumstances, one might expect to find more than ten thousand galaxies, more or less, in a volume that size, so the absence of those galaxies is striking. Indeed, the greatest surprise about voids is not so much their existence as the question of how their existence could have eluded the astronomical community until 1981, when the discovery of the first void, in the constellation Bootes (the Herdsman), was announced.

The answer to this puzzle has to do with a fundamental fact about the way the universe is observed. When we take a photograph of the night sky through a telescope, what we see is a two-dimensional image. Each star and galaxy shows up as a bright spot on the photograph, no matter how far away it is. All that matters is how much light strikes the negative.*

The importance of the two-dimensional character of our observations will be readily understood if you consider the situation shown in Fig. 5.3 (page 76). Suppose there are two sets of lights on the ground somewhere — lights separated by a distance of several yards — and suppose you are standing some distance from those lights with a camera. What would a picture show?

Obviously, you'd see something like what is shown on the right in the illustration. The planes on which the two lights shine would overlap, and you'd see a single, uniform display of lights in the photograph. There would be no hint whatsoever of a gap between the two planes; the camera combines them when it makes up the picture. The only way such a gap could be detected would be if some information about the third dimension — the distance between the two sets of lights and the camera — could be obtained. In the case of the lights on the ground, such information might be difficult to obtain, but in the case of the galaxies it is not.

In Chapter 2, we saw that the Hubble expansion was such that the farther away from us a galaxy is, the faster is its recessional velocity, and the greater the redshift of the light we receive from it. The redshift, then, provides us with a means by which the distance to the galaxy — the third dimension, in the analogy — can be estimated. This means that there is a way to

---

* Let me add that although I use here the language of photography to make the description easier to follow, almost no major telescopic observations these days use photographs. Electronic systems similar to television cameras have long been in use instead. This technical difference does not change the thrust of the discussion.

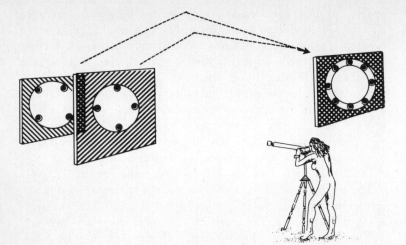

FIGURE 5.3

obtain a full three-dimensional picture of the entire cosmos and determine once and for all what the large-scale structure is.

Unfortunately, while this conclusion is correct in principle, in practice things are not so simple. To get the usual two-dimensional picture of the universe, all that is needed is to point a telescope, expose a photographic emulsion, and develop the image. To get a three-dimensional image, on the other hand, we must somehow look at each of the hundreds (or even thousands) of galaxies in our field of view, record enough of the light that comes from each so as to make an accurate determination of its redshift, and then put all of our observations together with the aid of a computer. Only then can we come up with the final image. The collection of data gathered in this way is called a redshift survey. It is not the work of a single night. It requires months and even years of work by a dedicated team of professionals.

Our earlier surprise, then, that the discovery of large-scale structure from these redshift surveys was not made until 1981 was unjustified; the fact really isn't surprising at all. There is just too much work involved in doing redshift surveys of the entire sky. The team of astronomers from the University of

Michigan who announced the result of a survey of galaxies in the constellation Bootes chose to "punch a hole in the heavens" by doing a very detailed survey of a very small region instead of a full sky map. The result of their work was the discovery of the largest void in the universe.

The bubble in Bootes is a volume some 250 million light-years across, and it appears to contain no normal galaxies whatsoever. It may contain a few dwarf galaxies, but that scarcely matters. The average distance between galaxies elsewhere in the universe is a matter of a few million light-years; that so few galaxies should exist in a space so large is truly astounding.

During the last few years, cosmologists have started to refer to voids informally as "Hubble bubbles." I like this name—it both captures the free spirit of modern cosmology and honors the man whose work started it all (although, to be honest, I doubt if Hubble would have appreciated an honor bestowed in this form).

The announcement of the Bootes void caused quite a stir in the science press, but made less of an impact among cosmologists, because of a characteristic of scientists that the public is hardly aware of—scientists often greet novelty without excitement, because a single discovery can always be a fluke. Hence it is usually safe to ignore it, at least until it is confirmed by an independent experiment.

As far as the void in Bootes is concerned, its existence could always be attributed to chance. After all, if the galaxies are distributed randomly in space, there are bound to be a few empty regions, just as there will occasionally be empty spaces in a large crowd of people gathered in an open plaza. I know that when I heard about the void in Bootes, I filed it under the "gee, that's interesting" category and went on with my work. I had been burned too many times by phenomena that turned up once or twice and were then never heard of again to fall into that trap another time.

But in the fall of 1985, the other shoe fell. A group of astronomers at the Harvard-Smithsonian Center for Astrophysics in Cambridge, Massachusetts, announced the result of *their* redshift survey. They had "punched a hole in the sky" in a completely different direction from the Michigan group, but they

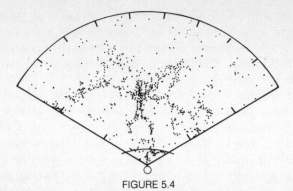

FIGURE 5.4

had found the same thing—the universe was full of large bubbles. With this announcement, the Hubble bubbles came into their own in cosmology. You can ignore one bubble as a possible statistical fluke, but the chance that two deep-sky redshift surveys, looking in different directions, would both just happen to find voids if the voids were really rare is infinitesimal.

I particularly enjoyed the fact that the map of the Cambridge survey had the form sketched in Fig. 5.4. Can you see a person in it? A little anthropomorphism never did anyone any harm. In the map, the earth is at the point of the wedge, and the distance from the earth increases as we move away from the point toward the wide part. In effect, the wedge is a slice through a piece of the universe. Each point in the diagram represents a single galaxy that has been located by using the three-dimensional techniques described above. In the period since the publication of this result, other redshift surveys have been done, and all seem to show evidence for the existence of voids.

The Spongy Universe

The picture of the universe that emerges from these studies is a striking one. Galaxies are not scattered uniformly through the universe, and neither are they scattered randomly. Instead a cross section of the universe looks like what you'd get if you sliced through a sponge. The solid matter would be arranged in an interconnected, stringy network interspersed with large bubbles in which no (or very little) matter is seen. Any attempt to explain the structure of the universe must confront this new vision of the way matter is arranged. How did it get that way?

In the previous chapter, we discussed some of the problems that one encounters in trying to explain the way that matter is clumped on a large scale. The existence of voids makes the problem even more difficult. There are two general classes of answers to questions of this kind: those that involve events that happened fairly late in the history of the universe, and those that involve the survival of structures formed during the first fraction of a second. In essence, the former class of answers would have it that galaxies formed first and were then cleared out from certain regions, leaving the bubbles behind. The latter class would say that galaxies formed on the edges of voids, where we see them, and for some reason did not form in the voids themselves.

Before going further, I must issue a couple of warnings. Hubble bubbles are a fairly new (and hot) topic in cosmology. Ideas about them are flying around in a heady atmosphere typical of the frontiers of science. This means that someone will think of something that might work and publish it, only to have someone else show that the notion has other, unanticipated consequences that cannot be reconciled with observation. That there should be second thoughts about new theories is usual, but when it happens in science some people find it hard to swallow. They seem to think that once a theory is promulgated, it possesses some sort of eternal truth. Actually, a theory is nothing more than a guess about the way nature behaves, and it is not accepted by scientists until all its consequences have been worked out and thoroughly tested. Many of the wild ideas that hit the press die quickly, and people are naturally a little bewildered. Where, for example, are the "parallel universes" of yesteryear?*

As I write this in the spring of 1988, the ideas discussed below are the center of attention in the cosmological community. By

---

*This refers to an idea that grew out of the Grand Unified Theories (see page 48) in the early 1980s. The version of the theories then being considered seemed to predict that the cosmos was made up of a foam of self-contained universes like our own. Unfortunately, after the theory was touted in *Time* and other publications, it died.

the time you read my words, they may have gone the way of the parallel universes. You should regard them, therefore, as nothing more than examples of theories that are being tested. They certainly are not the final answers to the problem of the voids and large-scale structure.

In one sense, it's not too surprising that the universe should resemble Swiss cheese. You can't make a pile of dirt without digging a hole, and by analogy, if you're going to have regions where galaxies cluster, you might expect to have regions where they are rare. Early on, a group of astrophysicists at Princeton proposed a very reasonable physical explanation of the voids. They suggested that after the galaxies had formed, there was some sort of titanic explosion in a region of space. The exact nature of the explosion, as well as its cause, varies from one researcher to the next. Whatever the source, however, an explosion could easily cause a massive shock wave that would sweep through a region of space, pushing to the outside any material that was there and producing the kind of bubble walls shown in Fig. 5.4.

That is an attractive idea, and it certainly provides an intuitively believable solution to the problem. The bubbles do look like what might result from a series of explosions in a more or less uniform medium. Unfortunately, if a mechanism could be found to produce such large explosions, there would be a conflict with the measurement of the cosmic microwave background (see Chapter 3). A large explosion may be considered a "fire" in itself, so that we should expect to see the radiation emitted by the cooling explosion, just as we see the microwave background resulting from the Big Bang. Calculations indicate that any explosion large enough to produce a Hubble bubble would also produce enough radiation to skew the cosmic background. Since such skewing is not seen, the argument goes, the bubbles cannot have been caused by explosions.

To counter the argument, the defenders of the explosion hypothesis point out that a bubble need not be produced all at once. It could arise from the coalescing of several smaller bubbles, like soap lather in your bath, foam in your beer (or champagne, of course), or bubbles you blow for your children. Whether this argument will stick remains to be seen, but the explosion idea is a good example of an explanation for voids that depends on something that happened after galaxies had already formed.

By contrast, other theories start from the premise that the voids and superclusters resulted from events set in motion long before the galaxies condensed out of the primordial gas. These theories assume that the mass concentrations around which galaxies formed were not uniformly distributed in space, but showed from the start the Swiss-cheese structure that we see in our surveys. On such a supposition, the galaxies would form *in situ* around the edges of the bubbles and would stay there.

For example, one type of theory that is enjoying a certain vogue involves something called a cosmic string (see Chapter 12). This is a long, very dense structure produced at 10^{-35} seconds that could easily be a condensation center for galaxies. If the universe was full of strings, then galaxies would form along them, producing superclusters. In this sort of model, the voids would be the spaces between the strands.

But all this is speculation. What has become clear is that explaining voids and large-scale structure in the universe is going to be a very difficult task for cosmology.

SIX

Dark Matter
Less Than Meets the Eye

All modern experiments tend to explode older theories.

—JULES VERNE,
A Voyage to the Center of the Earth

S O WE CANNOT EXPLAIN why matter in the universe is clumped
into galaxies, nor can we explain why those galaxies are
found in voids and superclusters. It would seem that the
last thing we need at this stage is another mystery. Yet anyone
who has ever put a puzzle together knows that adding one more
piece sometimes allows us to see the entire pattern, a pattern
that eluded us until the final piece was available. The discovery
of what has come to be called dark matter plays exactly this
role in the puzzle of the universe.

The reason for this state of affairs is simple: Up to this point
we have discussed the problem of explaining the universe under
the assumption that we were dealing only with visible matter
—matter that we see when we look through a telescope, or at
least perceive with radio telescopes and other devices that are
sensitive to wavelengths outside the range to which our eyes can

respond. If, as we shall argue presently, much of the matter in the universe is not in this familiar form, then all of the arguments about "time windows" and "running starts" have to be rethought. The properties of the new forms of matter, after all, may be quite different from the properties of the matter that we see in our laboratories, and these unexpected properties may allow us to get out of some of the dead ends we have encountered.

The notion that there is matter that cannot itself be seen but can nevertheless affect what is seen is not easy to accept. To get used to the idea that this might be so, we can start by looking at our own neighborhood and our own galaxy, the Milky Way.

The Milky Way—A Typical Spiral Galaxy

We can't see our galaxy from the outside, but we can see enough other galaxies to get a pretty good idea of the structure of the Milky Way. Well over half the galaxies in the universe have the same general shape, with a bright central core and two (but occasionally more) spiral arms. The spiral arms are certainly the most striking feature of galaxies, and I suspect that if the average person were asked to draw a picture of one, he or she would produce something very much like Fig. 6.1.

FIGURE 6.1

The existence of spiral arms tells us two very interesting things about galaxies. First, they rotate, and second, the most visible spots in galaxies are not necessarily the places where most of the matter is located. That these points follow from the spiral structure isn't immediately obvious, so we will make a slight digression to discuss them.

When you stir cream into your coffee, you often see transitory spiral patterns in your cup. The reason for these spirals in the cream is what physicists call differential rotation. The fluid at the edge of the cup is slowed down by friction and tends to stick, while the fluid in the center of the cup flows freely. Consequently a series of points that form a straight line at one instant of time will be carried different distances by the fluid in the next instant. Thus, a straight line will be rapidly transformed into a spiral, as shown in Fig. 6.2. This is what you see in your morning coffee, and it's tempting to jump to the conclusion that your morning spiral and the spiral of the Milky Way are somehow connected.

Unfortunately, this conclusion turns out to be wrong, as you will realize if you think about a simple observational fact. The sun is located about a third of the way out from the center of the galaxy. As the galaxy rotates, the sun and its planets are being carried around with a velocity of about 250 kilometers per second. At this rate, the sun would have had a chance to

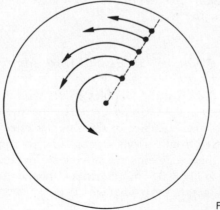

FIGURE 6.2

complete roughly sixty complete circuits since the Milky Way was formed. If the spiral arms were like the cream in your coffee—straight lines of dense matter that have gotten distorted into a spiral shape because of differential rotation—then they would long ago have become "wrapped up" and ceased to exist.

Modern thinking is that the spiral arms in galaxies stand out because they are regions where new stars are flaring into existence, and are therefore brighter than their surroundings. We notice them for the same reason that we notice the downtown area of a city from an airplane; both are emitting more light than what's around them. And just as it doesn't necessarily follow that the most brightly lit areas of a city have the highest population, neither does it follow that most of the matter in galaxies is in the spiral arms.

In point of fact, the matter in the galaxy is spread reasonably uniformly throughout the disk. The spiral arms stand out only because they are bright, not because they contain more material. This leads us to one of the most important ideas that we will discuss in this book, the idea that *there is no necessary connection between the presence of matter in a region and the emission of light or other radiation from that region.*

There are many ways in which matter can make its presence known, and emitting light (or radio waves or X rays or what-have-you) is only one of them. In fact, the only thing that all matter *must* do is exert a gravitational force. The final test for the presence of matter, then, is not whether it shines, but whether it attracts other matter to it.

Galactic Rotation Curves and Dark Matter

In Chapter 2 we discussed the Doppler effect. We saw that the perceived frequency of light can be altered by the motion of the source. Because of the existence of the Doppler effect, we can look at light emitted from various sections of a rotating galaxy and tell how fast these segments are moving in the plane of that galaxy, as shown in Fig. 6.3. The velocities obtained in this way, when plotted on a graph, form what we call the galactic rotation curve.

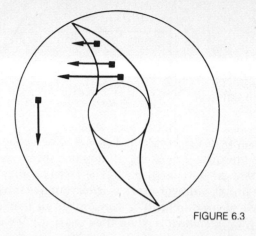

FIGURE 6.3

There are many possible shapes a rotation curve could have, all of which have analogues in our everyday experience. For example, when you're riding on a merry-go-round, you know you get a lot dizzier if you're on the outer edge than on the inside. That is why you start little kids off on the inner horses and let them work their way outward as they get older. The reason for this phenomenon is simple: The merry-go-round is a solid piece of construction, and when it turns the outside has to move faster to stay in line with an inner point. This sort of motion gives rise to a rotation curve like the one shown on the left in Fig. 6.4 (page 88) — a curve in which the velocity increases as we get farther away from the center. This is called "wheel flow" in the jargon of physicists, for obvious reasons. We would expect to find it whenever material is locked tightly together, as it is in a merry-go-round.

You can picture another common type of flow by thinking of a group of runners on a track, each in one of the lanes stretching from the inside to the outside of a running surface. Suppose the runners are all equally skillful, so that they all move at the same speed. If there is a curve in the track, then the runners' line will start to bend, with the inner runners moving ahead because they have to traverse less distance. If you measured a rotation curve for the runners, you'd get something like

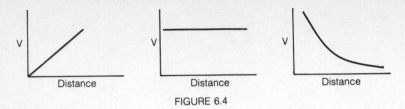

FIGURE 6.4

the one shown in the center in Fig. 6.4. The runners will all cross your imaginary line going at the same speed, so the curve will be a straight horizontal line. (The fact that they cross the imaginary line at different times is irrelevant as far as this particular type of measurement is concerned; all that matters is how fast they are moving when they cross it.) We'll call this situation "constant-velocity flow." It will occur, along with the characteristic horizontal rotation curve, whenever things all move at the same speed regardless of their distance from the center. Constant-velocity flow necessarily leads to differential rotation — the sort of thing we saw in the coffee cup — because the outer points, while they have farther to go, are constrained to move at the same speed as their inner counterparts.

Incidentally, it is the effect we are discussing here that is responsible for the fact that runners are given staggered starts in track competitions. The offsetting of the starting positions is designed to compensate for the differences in distances measured along the inside and the outside of the track.

A third type of rotation curve can be understood by thinking about the planets in the solar system. It is well known that the length of the "year" — the time to complete one orbit — is different for the different planets. It ranges from eighty-eight days for Mercury to almost 250 years for Pluto. To some extent this is attributable to the fact that the outer planets have farther to go to make a complete circuit, but that's only part of the story. It also turns out that the farther away from the sun a planet is, the more slowly it moves. A rotation curve for a system like the planets is shown on the right in Fig. 6.4.*

A curve of this type is called Keplerian, after Johannes Kepler

---

* The mathematical form of the curve is 1/r.

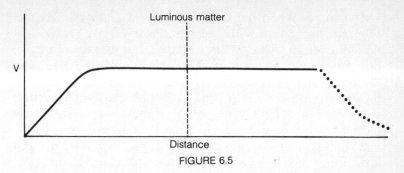

FIGURE 6.5

(1571–1630), the man who first worked out the correct laws for the orbits of the planets. We expect to find it in any situation where the mass that exerts the force of gravity lies at the center of the system, as it does in the case of the sun and the planets. It is important to realize that this requirement does not mean that the central mass has to be small. The sun is certainly not a small object. The solar system is Keplerian because the size of the sun is small compared to the distances to the planets. By the same token, if we get far enough away from the main concentrations of mass in the galaxy, the rotation curve should definitely be Keplerian.

This is a crucial point, so let me elaborate on it a bit. Suppose for the sake of argument that all the matter in the galaxy were concentrated in a sphere 100,000 light-years across—that it coincided, in other words, with the distribution of luminous matter. Then if there were some satellites (suns, say, or even single planets) in orbit 200,000 or 300,000 light-years away, we would expect their rotation curves to be Keplerian. The farther out such objects were, the slower they ought to move.

A typical measured rotation curve for a galaxy is shown in Fig. 6.5. Toward the nucleus, where the matter is tightly packed, we see velocities rising with increasing distance: wheel flow. Farther out, the curve levels off and we're into a regime where everything moves with about the same velocity, and hence the kind of twisting associated with differential rotation appears. This part of the curve extends well out past 100,000 light-years, and hence goes beyond the region that we can actually see when we look at a galaxy.

You may be wondering how we can know about the behavior of matter beyond the visible region. Obviously, no visible light reaches us from whatever lies out there, but that doesn't mean that there's no radiation at all. There are thin wisps of hydrogen gas in this region, gas that emits radio waves that can be detected by receivers on earth. The kind of Doppler analysis described above for visible light can be done on radio waves, so we can indeed find out how fast the hydrogen is moving.

If the gas is in a region of empty space, it will be in orbit around the galaxy—a species of atomic satellite. In this case, we should see the rotation curve turn over into the Keplerian form. If, on the other hand, the gas is being carried along by unseen matter like flotsam on the surface of a stream, then the rotation curve of the hydrogen will be the same as that of the unseen matter in which it is embedded.

As you can see in Fig. 6.5, the rotation curve of a typical galaxy remains flat well outside the region where visible light is emitted. In fact, as of this writing, *no* galactic rotation curve has ever been observed to turn over and become Keplerian. All of them remain flat out to distances of 200,000 or 300,000 light-years—several times the radius of the visible part of the galaxy. This is true of the many galaxies for which measurements have been made, and hence it is most likely true for the Milky Way as well.

We know that as soon as we get to a point where we are outside most of the matter in the galaxy, the rotation curve will become Keplerian and start to drop. The fact that it is not observed to do so can mean only one thing: Even at large distances from the center, distances well outside the boundary of the visible galaxy, there are appreciable amounts of matter. We may not be able to see it, but we know it's there because of the gravitational effects it has on the rotation curve.

We can, in fact, estimate the amount of extra matter in the Milky Way and other galaxies by studying the orbits of the galactic satellites. How much mass would it take to produce the observed rotation curves? The answer is that there must be at least ten times as much unseen matter in the galaxies as there is visible matter. In other words, at least 90 percent of the matter

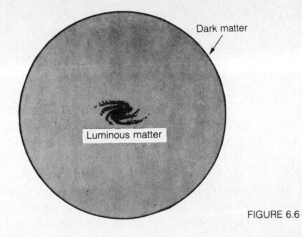

FIGURE 6.6

in galaxies like ours is in a form which does not give off visible light or other radiation, and whose very existence wasn't even suspected until the 1970s. Appropriately enough, astronomers have taken to calling this stuff "dark matter." It is spread throughout a large spherical halo surrounding the visible regions of the galaxy (see Fig. 6.6).

Dark Matter: Some Philosophical Comments

We will be discussing the consequences of the existence of dark matter and current thinking about what form it might take throughout the rest of the book, but it would probably be a good idea to pause briefly at this point and put its discovery into perspective. At first glance it may appear that dark matter is just one more little piece of the jigsaw puzzle that we have to assemble to understand our universe, neither more nor less important than many others. This would be a reasonable point of view if dark matter made up only a small part of the universe. In that case, we could regard its existence as little more than a footnote to the more important (and more easily detected) luminous matter.

But this point of view, reinforced as it is by our understandable prejudice that what we can see must be more important than what we can't, stands the universe on its head. In fact, the dominant form of matter is not luminous, but dark. It would not be too much to say that since over 90 percent of the matter in the universe is dark, the luminous matter — the stuff we actually see — may be no more important than debris on the surface of a stream. It may be that the bright spirals of the galaxies simply serve as passive markers, mute testimonials to forces operating at a level invisible to us. It may be that when we know more about our universe, we will recognize that our hard-won knowledge of the visible universe is little more than the first step on the road to an understanding of the way things really are. Many of the modern theories discussed in Chapters 11 and 12 have already adopted this point of view, but so far this new attitude has been confined to a small circle of experts.

In Chapter 1 we charted the evolution of our ideas about the universe from the earth-centered cosmology of the Greeks to the sun-centered world of Copernicus. We saw how this dethroning of the earth, its movement from the center of the universe to a role as just one more planet, had the effect of destroying the comfortable medieval world.

The dethroning of the earth continued with the work of Shapley and Hubble. Shapley showed that the sun was not even at the center of the Milky Way, but somewhere out in the spiral arms. Then Hubble showed that the Milky Way itself was just one of an infinite number of galaxies in an expanding universe.

But as familiar as Hubble's universe may have become during the past sixty years, the discovery of dark matter leads to profound changes in the way we perceive it. It may be that this familiar universe is itself little more than a minor part of the true working of things. It may even be that the kind of matter that makes up our solar system, our earth, and our bodies is itself a relatively minor part of a universe that is composed predominantly of very different stuff. It's hard to imagine a more gloomy ending to the Copernican odyssey, particularly if you want to believe in your heart of hearts that the earth and its human inhabitants occupy a special place in the universe.

Galactic Clusters: More Dark Matter

Rotation curves are one important indication of the existence of dark matter, but they aren't the only ones. There is more dark matter out there than is found in galactic halos, a fact that can be made clear by thinking about galactic clusters.

What is interesting from the point of view of dark matter is that there are many clusters that contain thousands of galaxies. In these clusters the galaxies move in complicated paths. In fact, you can think of the galactic cluster as being analogous to a drop of water suspended in space. Each of these systems is made up of constituents (galaxies for one, molecules for the other) that are free to move around. In addition, in each system there is a force (gravity for one, surface tension for the other) that keeps the constituents from flying off.

In the case of the water drop, we know that if we raise the temperature, the velocity of the molecules will increase until eventually they are able to overcome the forces of surface tension. When this happens, we say that the drop boils away. You could, in fact, predict whether the drop would boil or not by measuring how fast individual molecules were moving.

In just the same way, astronomers can measure the velocities of galaxies in a distant cluster and tell whether or not those galaxies could overcome the force of gravity exerted by other members of the cluster. The force of gravity, of course, depends on the amount of matter (both luminous and dark) that resides in the galaxies in the cluster. When these kinds of velocity measurements are made, a rather startling fact emerges. It turns out that in almost every case the velocities of the individual galaxies are high enough to allow them to escape from the cluster. In effect, the clusters are "boiling." This statement is certainly true if we assume that the only gravitational force present is that exerted by visible matter, but it is true even if we assume that every galaxy in the cluster, like the Milky Way, is surrounded by a halo of dark matter that contains 90 percent of the mass of the galaxy.

There are two possible ways to explain this result. One is to take it at face value and say that the cluster is indeed boiling away, but that we just happened to look at it while it was in the process. This would be like seeing a drop of water on a hot skillet before it had completely evaporated. This argument would be a reasonable explanation for the velocities in a few galactic clusters, but when you see the same pattern repeated everywhere, you begin to wonder. How likely is it that we came along just at the time when half the galaxies in the universe happened to be in clusters that were too weakly bound to survive and were therefore in the process of flying apart?

The alternative is to assume that the galaxies have always been grouped into clusters, more or less as they are now, and that the forces holding the clusters together are stronger than what we expect based on the amount of matter in the galaxies alone. One way that this could happen is if there is extra dark matter, over and above the galactic halos. Such matter could be secreted in the empty spaces between galaxies in the cluster. If this were the case, then the extra dark matter would exert an extra gravitational force on the galaxies, which would then be locked into the cluster despite their high velocities. This is the interpretation of the galactic-cluster data that is now accepted by most astronomers.

It appears, then, that dark matter appears in the universe in more than one place. It appears around individual galaxies, but it also appears between galaxies in clusters. In the jargon of cosmologists, we say that dark matter appears on many scales. Indeed, we shall argue that dark matter appears on *all* scales in the universe, and that wherever we see luminous matter, we should expect to find dark matter as well.

Having made this point, let us now turn to the question of how the existence of this heretofore unexpected type of matter can solve the problems we've encountered in trying to explain the structure of the universe.

SEVEN

How Dark Matter Could Solve the Structure Problem of the Universe

Now entertain conjecture of a time
When creeping murmur and the poring dark
Fills the wide vessel of the universe.

— WILLIAM SHAKESPEARE
Henry V, Act IV, Sc. 1

THE DISCOVERY OF dark matter reveals that most of the universe is composed of stuff we can't see. It is only natural to ask what effect this realization must have on the problem of explaining large-scale structure. It may seem strange that cosmologists should pin their hopes of understanding the universe on such mysterious stuff as dark matter, but that's exactly what is happening today.

Nor is this simply a case of grasping at straws — of taking advantage of our ignorance of the nature of dark matter to assign to it whatever properties are needed to solve the problems at hand. In fact, we shall see that we really don't need to know the details of the way dark matter behaves in order to see how it could solve the problem of galaxy formation. With the recognition of dark matter, we seem to have the final piece that we needed to put together the jigsaw puzzle and complete the picture of the way the universe came to be the way it is.

The basic idea as to the role of dark matter is simple to understand. As we saw, the main difficulty in picturing how the universe evolved has to do with the fact that if the entire cosmos is made up of normal matter, galaxy formation can't start until fairly late in the game, after the universe has cooled to the point where atoms can exist and radiation can decouple. By that time, the Hubble expansion would have spread matter out so thinly that gravity alone would not be strong enough to pull clumps together before everything got out of range. In Chapter 4, we also saw that all the ways we could think of to give the process a "running start" seemed to contradict some known observational fact.

Once we accept the idea that most of the universe is not in the form of familiar matter, this difficulty loses some of its power. Although the pressure of radiation interacting with protons and electrons in the plasma of the early universe may indeed prevent the clumping of ordinary matter until after atoms are formed, there is no reason whatsoever why the same should be true of dark matter. Suppose for the sake of argument that we had a candidate for dark matter that stopped interacting with radiation very early on in the Big Bang—during the first second, for example. This situation could arise if the interaction of the dark-matter particles with radiations depended on the energy of the collisions between the two and hence became small once the temperature fell below a certain level. In such a case, the dark matter could start to come together into clumps under the influence of gravity long before the formation of atoms. Radiation pressure would not prevent this sort of clumping, because our hypothesis is that radiation couldn't push on the dark matter as it does on ordinary matter.

If this happened, then when atoms finally formed and normal matter was free to begin aggregating, it would find itself in a universe in which enormous concentrations of mass already existed. Bits of ordinary matter would be strongly attracted to the places where the dark matter had already congregated and would move quickly to those spots. The process would be like pouring water onto a surface pitted by deep holes—the water would quickly run into the holes and the speed of the runoff would have almost nothing to do with the force that one bit of water exerted on another. So, the argument goes, once ordinary matter is freed from the restraints imposed by the pressure of radiation, it will fall into the "holes" already created by dark matter, and thus galaxies and other structures will form very quickly after radiation decouples. All the arguments about the "time window" that caused us so much grief in Chapter 4 would become irrelevant.

The beauty of this idea is that it takes two problems — the inadequate time window for galaxy formation and the existence of dark matter — and puts them together to form a solution to the central problem of the structure of the universe. The dark matter, by hypothesis, has a much longer time window than ordinary matter, because it decouples earlier in the Big Bang. It has plenty of time to clump together before ordinary matter is free to do the same. The fact that ordinary matter then falls into the gravitational hole created in this way serves to explain why it is that we find galaxies surrounded by a halo of dark matter. The hypothesis kills two birds with one stone.

But we must bear in mind that at this point we have only a notion that *might* work, not a well-thought-out theory. To go from the notion to the theory, we have to answer two important and difficult questions: (1) *How* does the dark matter explain structure? (2) *What is* the dark matter?

Hot and Cold Dark Matter

We can start examining these questions by thinking about the way that dark matter could separate from the hot, expanding cloud of stuff that constituted the early universe. By analogy with the discussion of the decoupling of ordinary matter after the formation of atoms given in Chapter 3, we will call the separation of dark matter "decoupling" as well. A transformation such as that leading to the formation of atoms is not needed for decoupling to occur. All that has to happen is that the strength of the interaction of the particles making up the dark matter fall below the point where the rest of the universe can exert a reasonable pressure on it. After that, the dark matter will carry on in its own way, indifferent to everything else that goes on around it.

It turns out that from the point of view of creating the observed structure of the universe, the most important characteristic of the decoupling process for dark matter is the speed of the particles when they are freed. If decoupling occurs very early in the Big Bang, the dark matter may come off with its particles moving very fast—almost at the speed of light. If this is so, we say that the dark matter is *hot*. If the decoupling occurs when the particles are moving slowly—significantly less than the speed of light—we say the dark matter is *cold*.*

Of the types of dark matter cosmologists have under consideration, the neutrinos (see Chapter 10) are the best example of hot dark matter. All the other particles we'll discuss in Chapter 9 are cold, and one type of dark matter—cosmic strings, discussed in Chapter 12—doesn't fit into this scheme of classification, because it isn't made up of particles at all. In the rest of this chapter we shall look at how hot and cold dark matter would function in an expanding universe and defer the question of what the dark matter is till later.

It turns out that hot dark matter acting alone almost certainly could not explain what we observe in the universe and that the cold-dark-matter scenario has to be modified extensively if it is to be kept alive as a candidate for the ultimate theory. It is this unsatisfactory state of affairs, together with the uncertainty about what kind of particles would constitute cold dark matter, that eventually led theorists to consider cosmic strings.

Free Streaming and the Demise of Hot Dark Matter

When cosmologists talk about hot dark matter, they have in mind (as we just saw) a particle called the neutrino. Its properties will occupy us in more detail as we go on. For the moment we simply note that it is a particle emitted in nuclear reactions and

* Note that the terms "hot" and "cold" refer to the speed of the particles when they decouple, not to the temperature of the universe at that time. In principle, a light particle could be "hot" even if it came off late in the Big Bang, while a massive particle could be "cold" even if it came off early.

is therefore seen in our laboratories today. It has either zero mass or a mass that is very small, and it normally travels at (or almost at) the speed of light. Our present thought is that by the time the universe was a second old the neutrinos no longer were interacting enough with ordinary matter to be affected by the pressure of radiation. Thereafter, the neutrinos expanded and cooled on their own, creating a sort of mirror image of the cosmic microwave radiation.

One way of thinking about the process by which neutrinos decouple is to say that it occurs when a neutrino traveling through matter is unlikely to interact with that matter. Two factors affect this likelihood: the density of matter (which tells us how often neutrinos come near other particles), and the probability that a neutrino coming near another particle will actually interact. After the universe was one second old, this combined probability was low enough for us to say that the neutrinos had decoupled. Presumably the same sort of process could occur for any other type of hot dark matter.

To continue the story, as the neutrino cloud expanded, it moved through regions of normal density without much inter-action. But if it encountered a region of high density — what I called earlier a mass concentration — it would have interacted and blown the concentration apart. It would do so because the first probability mentioned above, that of encountering another particle, would be higher in the mass concentration than else-where. In the words of the Arizona astronomer Jack Burns, it would break up mass concentrations "much as a cannonball moving at high speed might scatter a loosely built wall without being appreciably slowed by the collision." This process would go on until expansion had cooled the neutrino cloud to the point where the neutrinos were moving at low velocities — less than a tenth of the speed of light, for example. At this level of energy they could no longer exert pressure even in very dense concen-trations and they would stop playing a role in gravitational collapse. During the time between decoupling and slowing down, the neutrinos are said to be "free-streaming."

Massive neutrinos, then, could break up mass concentrations for a time after decoupling. While doing this, they could not travel more than a finite distance — certainly no farther than light could travel in the same time span. In other words, if a neutrino starts its career in a very large concentration of matter, it will never get all the way through it in the time allotted. But if it starts in a small concentration, it will be able to get all the way through, acting as it goes like the cannonball in the loose wall. The upshot would be that free-streaming neutrinos would break up small mass concentrations, but would leave large ones more or less untouched.

Such a selective demolition of certain concentrations in the early universe, occurring long before radiation decoupled and gravitational collapse started in earnest, would mean that the neutrinos had destroyed every nucleus around which galaxies of less than a certain size could condense. That size can be estimated; it is computed from the distance traveled by neutrinos during free streaming. It turns out to be quite large—roughly the size of galactic superclusters. So, when gravitational collapse started, the smallest and fastest-growing centers of mass would have been the size of superclusters.

The existence of free streaming, then, tells us that hot-dark-matter theories must predict the following sequence of events: First the hot dark matter has to collect ordinary matter into clumps the size of superclusters, then these clumps have to break up into clumps the size of clusters, and then these have to break down into still-smaller, galaxy-sized clumps. This scheme is often called the "top-down" scenario for the formation of structure in the universe.

There is no mystery about how the breakup of superclusters into galaxies would occur: As we saw in Chapter 4, the ordinary workings of gravity are enough to break up any uniformly distributed material into small blobs. The problem is the times involved. Hot dark matter predicts a universe in which clusters of galaxies are old, but galaxies themselves are young. This is just the opposite of what we observe. The Milky Way, for example, contains stars at least fourteen billion years old—roughly the age of the universe. There are, in other words, stars in the Milky Way that were formed earlier than the hot-dark-matter models suggest the Milky Way itself could have formed.*

---

* For a discussion of how the ages of things like stars and universes are determined, see my book *Meditations at 10,000 Feet* (Scribners, 1986).

There are other difficulties with the massive neutrino as a candidate for dark matter—some will be discussed in Chapter 9. For the present, we chalk up the fact that hot dark matter involves such long free-streaming distances (and hence such large initial structures in the universe) that most cosmologists have abandoned it.

Cold Dark Matter and Biasing

Cold dark matter avoids this difficulty: The particles are moving so slowly when they decouple that they can't travel far during their free-streaming phase, and consequently even small concentrations of matter can survive. We have thereby a situation such that small groups of matter come together first, and these small aggregates gather to form the observed large-scale structure. This is called the "bottom-up" picture of the formation of the universe.

Furthermore, calculations done with the cold-dark-matter models show that the model offers a number of other successes. It predicts, for example, that galaxies would be created in a rather restricted mass range. The calculations indicate that cold dark matter should produce galaxies from about one-thousandth to about ten thousand times as massive as the Milky Way— none bigger or smaller. In point of fact, almost all known galaxies have masses within this range—a regularity that has always been a puzzle to astronomers. That this fact (along with several other details of the systematics of galaxies) can be explained simply by the cold-dark-matter hypothesis is regarded as a great triumph.

Unfortunately, the results of the redshift surveys and the discovery of the voids and filaments created serious objections to cold dark matter as the ultimate constituent of the structure of the universe. Even Marc Davis of the University of California at Berkeley, one of the strongest proponents of cold dark matter, wrote: "The cold dark matter model is now ruled out because it cannot produce voids as large as the one found in Bootes." The problem is that the small streaming distance of cold dark matter means that the universe must be built up from small, galaxy-sized objects, and it is hard to see how random collections of small objects could contain huge voids of the type that observers have found.

But clever ideas do not die easily. Davis and his co-workers at Berkeley began to think further about the relation between dark and luminous matter. They reasoned that when radiation decoupled, luminous matter would tend to be drawn to the largest surrounding concentrations of dark matter; that is, luminous matter wouldn't be scattered uniformly in space, but would tend to gather where large amounts of dark matter already existed. When we look out at the universe we are not seeing the regions where all the dark matter is, but only the places where it has pulled in enough luminous matter to create a galaxy or a galactic cluster. In current language, our view of the universe is necessarily "biased" because we see only luminous matter. The Berkeley group argue that it is quite possible that the dark matter is spread much more uniformly than luminous matter, so that the great voids we see may actually have dark matter in them.

Perhaps an analogy will help to make clear this "biased" view of the universe that presupposes cold dark matter as its basis. You know that the bottom of the ocean is full of small hills and mountains. Suppose that these elevations were scattered about more or less uniformly on the ocean floor, but that the floor itself had gentle undulations, as shown in Fig. 7.1.

If you could see the entire ocean floor, you would say that the material in it was distributed more or less randomly — there would be no signs of the sort of thing we have called large-scale structure. Suppose, however, that you could only see the land that stood up above sea level. Then, as shown, you would see

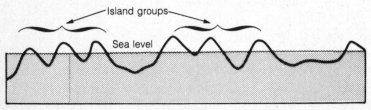

FIGURE 7.1

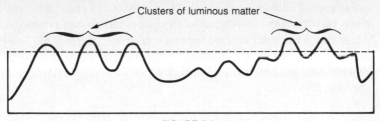

FIGURE 7.2

the sea broken up by a series of islands, but these islands would tend to cluster in regions where the high parts of the gentle undulations are found.* Looking only at the islands, you would conclude that undersea hills occurred in clusters and that the bottom of the ocean had a structure in which regions of high concentration of hills were interspersed with large voids — voids seen as large expanses of open water without any islands.

In exactly the same way, so the theorists argue, the voids and filaments we see in the sky do not trace out the true distribution of matter in the universe, but only the peaks and high points. They put this notion to work in the following way. They first calculate the distribution that they expect cold dark matter to have, getting something like what is shown in Fig. 7.2. Then they take only the peaks of this distribution — the parts that are above the dotted line — and say that those are the locations where we see luminous matter gathered into galaxies.

Using this line of attack, Davis and his colleagues have produced plots of galaxy distributions that look very much like what is actually seen in the sky. Whether this can be said to have solved the problem of the structure of the universe is another matter. The model based on the notion of bias still has trouble producing large voids with sharp edges — though the discovery of small galaxies in the Bootes void was a godsend to the theorists, because it relieved them of the need to get *all* the luminous matter out of their voids. My own impression is that, at the present time, cold dark matter with biasing cannot be ruled out as an explanation of the structure of the universe. Still, its proponents are spending a lot of time dodging bullets

* This is in fact the way that the Hawaiian Islands formed from undersea volcanoes.

while trying to patch up cracks in the theory. I should be surprised if it turned out to give us the last word on our problems. Marc Davis himself commented at a recent meeting that he "only believes it on alternate Wednesdays." Nevertheless, it remains one of our best bets for explaining what we have discovered in the sky.

Now What?

It seems, then, that dark matter plays an important role in the creation of large-scale structure in the universe, because it can decouple early and form mass concentrations to which luminous matter is latter attracted. If this is true, then some obvious questions arise. One, which we shall tackle in the next chapter, is how much dark matter there is in the universe. The second, already mentioned, is the question of what the dark matter is.

It appears that all we need in order to produce a universe with large-scale structure is matter made from particles that are massive (so that they can exert a gravitational force) and that move slowly when they decouple. Apart from these modest requirements, cosmology puts no restrictions on the form the particles of dark matter can take. This is a good thing, because it means that we have not tied our cosmological ideas too closely to a particular dark-matter candidate. At the same time, this lack of specifications limits our ability to determine what the dark matter is. In fact, we shall see that the best we can do is say that we don't know what makes up 90 percent of the universe, but we *do* know it's not something we've ever seen before.

EIGHT

Dark Matter and Missing Mass: How Much Should There Be?

> *Behold, a sower went forth to sow . . .*
> *And some seeds fell by the wayside,*
> *and the fowls came and devoured them:*
> *Some fell in stony places,*
> *where they had not much earth . . .*
> *And some fell among thorns*
> *and the thorns sprung up, and choked*
> *them:*
> *But others fell onto good ground*
> *and brought forth fruit . . .*
>
> —MATTHEW 13:3–9

YOU MIGHT THINK, on the surface of it, that the discovery of something like dark matter would set the astronomical community on its ear, cause all kinds of controversy and arguments between the Young Turks anxious to get on with new science and the Old Guard trying to defend traditional views. In point of fact, no such controversy ever erupted. Indeed, the discovery of dark matter appears to be one of the best-kept secrets in modern science. The great yawn that greeted the first announcements of the discovery can be attributed to the fact that most astronomers had believed for a long time that such material had to exist somewhere. They even gave the material a name—"missing mass"—before a single piece of evidence had been uncovered in favor of its existence.

The popular image of scientists is that they are a group of very hardheaded, "show me" individuals who believe something

105

only after it has been demonstrated to a fare-thee-well in the laboratory. In reality, "soft" considerations such as beauty, elegance, and simplicity play a much bigger role in the way scientists think than people suspect. If an idea isn't sweet — if there is not a brightening of the soul when it is explained — the proponents of the idea may have to accumulate overwhelming evidence before it will be accepted by the scientific mainstream.

Indeed, I have developed an easy rule of thumb to determine how tough a time a speaker is going to have selling his or her notions to the scientific community at large. While an idea is being proposed, say, in a talk, I simply look around the audience. If I see people break into a smile and nod when they catch on, I know there will be smooth sailing. Otherwise, look out! The acceptance of dark matter by the mainstream followed this pattern exactly.

When such an idea is adopted easily, it usually means that it's an idea that people want to accept, an idea for which the ground has been prepared. The expanding universe discovered by Hubble leads comfortably, as we shall shortly see, to the feeling that a universe in which much of the matter is not seen is more beautiful, more natural, than a universe in which everything is luminous. So when dark matter was discovered, the nods and smiles started and everyone said, "Of course — how could it be any other way?" It seemed so natural that no one bothered to tell the reporters about it.

The Spear of Archytas in an Expanding Universe

Remember the argument given by Archytas (discussed in the Prologue) about the infinite universe? Well, the only trouble with his reasoning was that he got the direction of the spear wrong. He wanted his spearman to walk to the edge of the universe and throw the spear out *horizontally*. What he had in mind (and what we think of when we read his arguments) is something like the situation shown on the left in Fig. 8.1. The spear is thrown in such a way that the force of gravity acts in a direction perpendicular to the direction of the throw.

FIGURE 8.1

In the Hubble universe, however, the spearman would en-counter a situation like the one shown on the right. The spear would be thrown outward, and the force of gravity would act in the same direction as the throw; that is, it would act in such a way as to slow the spear down and possibly pull it back into the known universe. Had Archytas lived today he would doubt-less realize that the best device to test the size of the universe involves throwing the spear *vertically*, letting it try to climb away from the earth while the force of gravity is trying to pull it back down.

The spear of Archytas for our age is a quasar like the one called QSO1208 + 1011, the most distant object ever observed in the universe. The spearman has thrown this object so that it is moving away from us (and from the bulk of the known universe) at almost the speed of light. The question to ask about QSO1208 + 1011 is the very question Archytas asked about his spear. Will it continue to fly outward, or will it turn around and come back at some time in the future?

If we imagined our spearman standing on the surface of the

earth instead of at the edge of the universe, the question would be easy to answer. We know how much mass is contained within the earth, and so we know exactly how great the gravitational force exerted on the spear will be. Whether the spear escapes into space or falls back to the ground depends on one thing only — how fast the spear can be made to move. If it goes faster than seven miles per second, it will escape. Otherwise, it will fall back; that's all there is to it. And though Archytas would have had trouble imagining a spear moving at that speed, the spearmen at NASA have had no difficulty at all in achieving it.

With distant quasars our difficulty is just the opposite of the spear escaping the earth. In the case of the spear, we know the mass of the earth and have to calculate the speed needed to escape. In the case of the quasar, we know the speed but not the size of the mass pulling it back.

Pondering the fate of the most distant quasar, one readily sees that the only force that could act to slow it down is the gravitational attraction of the rest of the mass in the universe. The total mass of the universe is the parallel to the total mass of the earth acting on the spear thrown from its surface. Both masses determine what the escape velocity of an object must be. Thus when we ask whether the quasar will ever turn around, we are really asking what the total mass of the universe is — in effect, we are asking: What does the universe weigh?

Since dark matter, like its luminous counterpart, can exert a gravitational force, it, too, will exert a force on that distant quasar. This point, as obvious as it is, must be made, because when we "weigh" the universe, we will have to put both dark and luminous matter in the scales if we are to obtain a reliable answer.

The force on a distant quasar bears on one of the most important questions we can ask about the Hubble universe: Will the expansion go on forever, or will it someday stop and reverse itself? This question fairly compels attention once the reality of the expanding universe has been established.

There are only three possible answers to the question, each corresponding to a different kind of universe: (1) There may not be enough matter in the universe to reverse the expansion. In that case the outward-moving quasars and galaxies will slow down as time goes by, but will never stop. We then say the universe is *open*. (2) There may be enough mass to slow down, stop, and reverse the motion of the most distant objects. The universal expansion we now observe will then be converted into a universal contraction, which some astronomers refer to (only half jokingly) as the Big Crunch. In this case, we say that the universe is *closed*. (3) The mass of the universe may be such that the gravitational pull is just enough to slow those outermost objects to a stop after an infinite amount of time has elapsed. In this case the expansion will slow down forever and come to a stop at infinity, but will never reverse itself. Such a universe is said to be *flat*. Of the three possibilities, as we shall see, it is the flat universe that is the most interesting.

Up to this point we have discussed the expansion in relation to gravitational forces slowing down distant galaxies. This is a mode of expression that is easy for us to understand, because it is similar to our everyday experience. It is not the language that an astrophysicist would use. In the scientific literature, it is usual to speak the language of general relativity. Another way of putting this is that up to this point we have used the language of Newton because it is more familiar. Now I want to digress briefly to introduce you to the language of Einstein, both because it's interesting in and of itself and because it makes some topics easier to understand.

"Open" and "Closed" in General Relativity

The first thing to know about general relativity is that at the conceptual level it requires no fearful effort to grasp. The popular idea of relativity as something only bearded scientists mumbling incomprehensibly can handle is a myth, nor is it true that only a dozen people in the world understand it. These impressions may have been plausible in the 1920s, but today they are simply silly. The basic concepts of relativity are routinely taught to liberal arts students in introductory astronomy and physics courses, and no one can go on to advanced undergraduate physics without having mastered them. Even the fully mathematical version of the theory is taught to hundreds of graduate students every year. I hope that by the end of this digression, it will be clear that understanding relativity is within the reach of anyone willing to put his or her mind to it.

Let's start with one of the more intimidating concepts, that of four-dimensional space-time. Like the man who was amazed to discover that he had been speaking prose all his life, most people are surprised to find that they have been using this concept all along. Try to recall the last time you said something like "I'll be in Chicago next Tuesday." That statement contains within it information that can be classified under the headings *When* and *Where*.

How many numbers would you need to specify this *When* and *Where*? First, you'd have to specify the location of Chicago. This would require three numbers. You might, for example, use latitude, longitude, and altitude to specify the point in space we call Chicago. In general, you need three numbers to specify any point in our normal three-dimensional space. You also need one number to specify the time implicit in your statement; that number might be the time on Tuesday when your plane will land. Thus, to give the meaning of the sentence with exactness, four numbers are required — three for spatial location and one for temporal. Taken together, these four numbers make a four-dimensional description.

In everyday life we do not usually think of time as a fourth dimension, because we assume that the river of time flows on independent of the place from which it is being observed; in other words, that time is the same in Nairobi or on Alpha Centauri as it is in Chicago. For everyday purposes, this assumption is more or less true — true enough, certainly, for the use we make of time. Clocks do not read differently when we go from one place to the next, nor do they change when we get into an airplane or car.*

But when we move at nearly the speed of light or when we're near very large masses, our everyday expectations about the independence of time and space no longer describe the universe correctly. In these unusual environments, the fourth dimension

---

* The changes associated with time zones, being merely man-made conventions, do not count here.

—time—becomes inextricably tangled up with the other three. Just as you can't go from Chicago to New York by a change of latitude and longitude alone (you must also change altitude), so when comparing motion between two fast-moving rocket ships you find a change not only in the spatial dimensions, but in the fourth dimension, time, as well. Clocks in the two rocket ships will tick at different rates.* This connectedness explains why in Newtonian physics we speak of "space and time" but in modern relativistic physics of "space-time."

We are now ready for another concept of relativity, that of "warped space-time." This is the most important idea to grasp for an understanding of the large-scale structure of the universe. But first, please note that there are two separate theories of relativity: special and general. Both are derived from a single basic principle, the principle of relativity. What it says is that *the laws of nature as seen by any observer are the same.*

If the observers are moving at constant velocities, then following the mathematical consequences of the principle leads us to the special theory of relativity. This theory, which was published by Einstein in 1905, contains most of the familiar unexpected effects of relativity—fast-moving clocks slowing down, objects getting heavier as they speed up, and so on. This theory has been well verified in many ways—indeed, the giant particle accelerators that push protons up to almost the speed of light are working examples of machines designed according to the precepts of special relativity. Their continued operation provides daily confirmation of the theory.

If we broaden our definition of "observer" to include even those observers who are accelerating, then following the principle leads us to the general theory of relativity, a much more difficult theory from a mathematical point of view. Einstein published it in 1915. It is general relativity that is our current best theory of gravity, and it is in general relativity that the concept of warped space-time arises. Skipping the mathematics, I'll now try to show you how acceleration is connected to gravity

---

* To be more precise, clocks will change when they move at any speed, but it is only near the speed of light that the change in running rate becomes important. A more complete discussion of general relativity is in my book *The Unexpected Vista* (Scribners, 1983).

through the principle of relativity, after which I'll suggest a simple way to visualize the universe as it is seen through Einstein's eyes.

You have probably had the experience of getting into an elevator on the ground floor of a tall building and feeling yourself being pushed down into the floor when you started up. Chances are you've also had the reverse experience when coming down from the top floor—feeling yourself about to float as the elevator started down. These feelings are not an illusion. If you were standing on a bathroom scale inside the elevator, you would actually see your "weight" swing up on the upward trip and down on the downward.*

The change in your apparent weight in a moving elevator is linked with the acceleration or deceleration of the cage. (You know this is true because you feel the change only when the elevator starts and stops.) This, in a nutshell, is why the general theory of relativity is a theory of gravitation. The principle of relativity tells us that every observer, accelerated or not, must see the same laws of physics operating in his or her frame of reference. If you think of the act of stepping on a scale and observing the reading as an experiment, the principle tells you that there is no way to determine whether the reading on the scale is due to the fact that you are standing on a gravitating body like the earth, or are in an accelerating body in the depths of interstellar space. In both situations, the scale will show weight. In short, the principle indicates that there is no experiment we could conduct that would tell us whether we were in an accelerating rocket ship or on the surface of a planet.

This link between acceleration and the effects of gravity is the cornerstone of the theory of general relativity. The mathematics of the theory are pretty difficult, but the results of starting from this connection and following it to its logical conclusion are easy to visualize, particularly with the aid of an analogy.

Imagine a sheet of flexible rubber marked out in a grid as on the left in Fig. 8.2. Then imagine dropping a large ball bearing

*If you are tempted to try this experiment, I suggest picking a time when there are only a few people around. I did it in the Sears Tower in Chicago, the world's tallest building, and got some very funny looks from my fellow passengers.

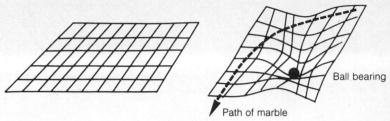

Ball bearing

Path of marble

FIGURE 8.2

on the sheet. The result will be a distorted sheet, as shown on the right. If you roll a marble across this distorted sheet, you will find that its path will be deflected when it hits the depression, as shown.

As in the elevator, there is no way you can tell whether the marble is deflected because the rubber sheet is deformed or because a gravitational force is acting between the marble and the ball bearing. In other words, there is no difference, as far as the motion of the marble is concerned, between a universe in which a gravitational force between bodies deflects the marble, and a universe in which the ball bearing distorts the sheet and that distortion, in turn, deflects the marble. The difference between these two ways of looking at the motion of the marble is, in essence, the difference between Newton and Einstein.

In Einstein's universe no forces act. What happens is that the presence of matter warps the fabric of space-time, just as the presence of the ball bearing warps the rubber sheet. We now regard motions that used to be attributed to the action of forces as due to distortions in the fabric of space-time induced by the presence of matter. It isn't that the universe of Einstein contradicts Newton's — rather, the same observed facts have been interpreted differently.

For example, we can reformulate the question of the open vs. closed universe discussed above in the language of general relativity. Imagine increasing the mass of the ball bearing in our example. Eventually we'll get a situation in which the hole caused by the weight will get deeper and deeper until it reaches the stage where it closes in on itself.

Now, imagine what will happen to a marble inside the self-

FIGURE 8.3

contained sphere created by the great mass of the ball bearing. You might push the marble out from the bottom, giving it a small velocity, in which case it would roll partway up the walls of the bubble and then fall back down. (See Fig. 8.3, left.) You might, on the other hand, imagine launching the marble from the bottom with a very high velocity, in which case it might go all the way around the loop. But no matter what velocity the marble has, sooner or later it will wind up back where it started. This amounts to an exact analogue of what we have called a closed universe. If the mass is big enough, in other words, you can take the Newtonian point of view and say that it exerts enough of a gravitational force to pull the marble back, or you can take Einstein's point of view and say that the mass is large enough to fold space back on itself ("close" it). Either way, you wind up with the same result.

If the mass of the ball bearing is just large enough, we can produce what we have called a flat universe. Far from the ball bearing the grid assumes more or less its undistorted shape. As in Fig. 8.3, right, it will form a flat surface on which the marble can roll (hence the origin of the term "flat universe").

When the mass is less than the amount needed to produce flatness, the universe is open. This situation really doesn't lend itself to being drawn on a two-dimensional page (we are, after all, dealing with four dimensions). I think of the open universe as being roughly analogous to the surface of a saddle—once a marble starts rolling it keeps going and never comes back.

As you can see, the similarities between the traditional and relativistic ways of looking at motion are many. This shouldn't be surprising—both describe the same gravitational force. When you come right down to it, there are only a few instances in which the differences between the two ways of looking at gravity can actually be measured. In general, relativity applies at very large distance scales and for very large masses, but gives the same results as Newtonian gravity for almost all situations where measurements can actually be made (i.e., for all phenomena that involve a distance scale smaller than a few million light-years).

This fact leads to one of the most surprising situations in the history of science. The general theory of relativity, unlike the special theory, has been tested rigorously in only two situations.* Despite the fact that it is one of the most revolutionary theories ever proposed, it has been accepted by scientists on grounds that are almost purely aesthetic. It is accepted because it is beautiful and elegant and therefore ought to be right. This acceptance is as good a piece of evidence as I can think of that scientific conclusions are often made by processes much more complex than a simple weighing of experimental facts.

In any case, it is clear that whichever language we use to describe the question of open vs. closed, the answer depends on one quantity: the total mass of the universe. About this quantity we can ask two different questions: How much mass is there, and how much mass should there be? Let's look at both questions briefly.

How Much Matter Do We See?

The total amount of matter in the universe is customarily given in terms of a quantity called the critical density, denoted by Ω_c. This is the density of matter needed to produce a flat universe. The actual density observed is then either less than or greater than this number. In the former case the universe is open, in the latter it is closed. The critical density isn't very large; it corresponds to about one proton per cubic meter of space. This may not seem like much, given the huge number of atoms in a

---

* The two tests of general relativity are the bending of light and radio waves as they come around the sun, and certain small effects on the orbit of Mercury. A third test, a measurement of the redshift of light as it climbs up out of the earth's gravitational well, tests the principle of relativity rather than the details of the theory.

cubic meter of dirt, but you have to remember that there's an awful lot of empty space out there between the galaxies.

Estimating the amount of luminous matter in the universe is a fairly easy thing to do. We know how bright the average star is, so we can get an estimate of the number of stars in a distant galaxy. We can then count the number of galaxies in a given volume of space and add up the masses we have found. Dividing the mass by the volume of the space gives us the average density of matter in that volume. When we carry out this operation, we find that the density of luminous matter is roughly 1 to 2 percent of the critical density—far below what is needed to close the universe. On the other hand, it's close enough to the critical value to give one pause. After all, this fraction could, in principle, have been one-billionth or one-trillionth, and it could equally well have happened that there was a million times as much luminous matter as is needed for closure. Why, of all the masses that the universe could have, is the measured mass of luminous matter as close as this to the critical value?

One could always argue that this is a cosmic accident—things "just happened" to turn out that way. I would have a hard time accepting this explanation, and I suspect others would, too. It is tempting to say that the universe actually has the critical mass, but that we somehow fail to see all of it.

As a result of this supposition, astronomers began to talk about "missing mass," by which they meant matter that would make up the difference between the observed and the critical densities. I never liked this term, because it carries a kind of religious overtone; when you say something is "missing," the implication is that it is supposed to be there. We have no proof that any mass is "missing," only a hunch that there might be more of the stuff hidden away out there.

This predisposition to believe in a more massive universe than is indicated by luminous matter is what I was talking about earlier when I said that the ground had been prepared for the existence of dark matter by the widespread belief in missing mass. To my great satisfaction (and, I must admit, surprise), the term "missing mass" has in recent years been dropped from the cosmological vocabulary. It has been replaced by the more accurate and more neutral "dark matter."

To get on with the story, the existence of galactic halos with masses ten times that of the luminous matter in a galaxy pushes the estimate of the density up to the region of 10 to 20 percent of the critical value. If we thought we were close to criticality with luminous matter only, we're certainly much closer now.

The dark matter in clusters and superclusters was discovered too recently for unanimity to prevail among astronomers as to its contribution to the total mass of the universe. A rather spirited debate on this subject is going on right now, but the bottom line seems to be that when this contribution to the dark matter is included, the total mass density of the universe is still no greater than 30 percent of the critical value.

Whether the value of this mass will continue to rise to the critical value as time goes on will be discussed in Chapter 13. For the moment, I simply quote a remark made by the British astrophysicist Steven Hawking. An observational colleague had been telling him that whatever dark matter there is had been found. Hawking replied, "Twenty years ago you had two percent. Today you have thirty percent. Why don't you go out and look again?"

What Should the Mass of the Universe Be?
The Role of Inflation

The idea of missing mass was introduced because the observed matter density in the universe is close to its critical value. Until the early 1980s, however, there was no firm theoretical reason to suppose that the universe actually had the critical mass. In 1981, Alan Guth, then at Stanford, now at MIT, published the first version of a theory that has since come to be known as the "inflationary universe." Since that time the theory has undergone a number of technical modifications, but the central points have not changed. For our purposes, the main feature of the inflationary universe is that it established, for the first time, a strong theoretical presumption that the mass of the universe must indeed have its critical value.

This prediction comes from theories that describe the freezing out of the strong force 10^{-35} second into the Big Bang. Among the many other processes going on at that time was a rapid expansion of the universe, a process that came to be known as inflation. It is the presence of inflation that leads to the prediction that the universe must be flat.

The process by which the strong force freezes out is an example of a phase change, similar in many ways to the freezing of water. When water changes into ice, it expands; a milk bottle will split if it's left outside overnight in winter. That the universe expands in the same way when it changes phase shouldn't be too surprising.

What *is* surprising is the sheer magnitude of the expansion. The size of the universe increased by a factor of no less than 10^{50}! This number is so huge that it is virtually without meaning to most people, including the author. Let me put it this way: If your height suddenly increased by a factor as large as this, you would stretch from one end of the universe to the other, with room to spare. Even a single proton in a single atom in your body, if its dimensions were increased by 10^{50}, would be larger than the universe. At 10^{-35} seconds, the universe went from something with a radius of curvature much smaller than the smallest elementary particle to something about the size of a grapefruit. No wonder the name "inflation" was attached to this process.

When I first read about the inflationary universe, I experienced difficulty with the rate of inflation. Didn't such a rapid growth violate Einstein's strictures against travel faster than light? If a material body traveled from one side of a grapefruit to the other in 10^{-35} second, its velocity would exceed the speed of light by a considerable amount.

The answer to this objection can be found in the bread-dough analogy in Chapter 3. During the period of inflation, it is space itself—the bread dough—that is expanding. No material body—none of the raisins—moves at high speeds within space. Strictures against faster-than-light travel apply only to motion *within* space, not to motion *of* space. Thus there are no contradictions here, even though at first glance there may appear to be. The consequences of the period of rapid expansion can best be described by reference to the Einstein view of gravitation. Before the universe was 10^{-35} second old, there was presumably

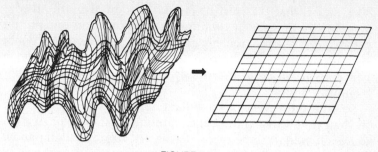

FIGURE 8.4

some sort of matter distribution (we'll see in a moment that its precise form makes no difference). Because of this matter, space-time would have some characteristic shape. For the sake of argument, suppose the squiggly* surface on the left in Fig 8.4 is taken as representing the configuration of the universe before inflation sets in.

You can imagine inflation as a stretching out of the sheet until its size exceeds that of the entire universe. No matter how convoluted the original surface was, when it is stretched out that far any section of it will look like the surface pictured on the right in Fig. 8.4. In other words, after inflation the universe will be flat *regardless of how it started out.* This flatness is no accident; it's a necessary consequence of the physics of the freezing that took place at 10^{-35} second.

When Guth hit the colloquium circuit with his theory, I can tell you there were a lot of smiles and nods. This solution to the problem of the missing mass (or, to use the cosmologists' term, the "flatness problem") is a perfect example of the elegance and beauty I mentioned earlier. The universe is flat because it can be no other way. Flatness is the only condition consistent with the basic laws that govern the interactions of elementary particles. No matter how the universe began, it will wind up flat. A truly sweet idea!

One surprising thing about inflation is how precisely flat it requires the universe to be. The observed mass of the universe has to equal the critical mass to an accuracy of one part in 10^{50}.

---

* Please excuse my use of this technical term.

This means that the level of tolerance allowed by the theory amounts to an error in counting of no more than one proton per cubic light-year — roughly one amoeba per galaxy. No doubt such accuracy is likely to remain outside the ability of human beings to achieve, but it certainly makes the prediction unambiguous.

So with the advent of the inflationary universe, we have for the first time a strong theoretical presumption in favor of a flat universe. Indeed, some theoreticians are already talking about the "requirement" of flatness. The fact that we now have 30 percent of the critical mass and expect with some measure of confidence to come to 100 percent makes the problem of discovering the identity of the dark matter all the more important. We shall now consider various ideas that deal with this puzzle.

NINE

Candidates for Dark Matter

Round up the usual suspects.

—Famous line from
the movie *Casablanca*

IT IS OBVIOUS THAT dark matter plays a very important role in both the evolution and the structure of the universe, but what sort of material is it? Can we really be satisfied that we understand the universe around us when the form of its most ubiquitous constituent — dark matter — is unknown?

Well, it is easier to say what dark matter isn't than to say what it is. But clearing the decks by negatives, though never a glamorous occupation, is often very useful. Consequently, we begin our considerations of the identity of dark matter by eliminating the most obvious candidates. At the end, we'll conclude that whatever dark matter is, it is something that has most likely never been seen on earth.

Baryonic Matter

The nuclei of the atoms that make up all the matter we encounter in our everyday lives are made primarily of protons and neutrons. These particles are members of a class that physicists call baryons ("heavy ones"). Consequently, you will often hear physicists referring to "baryonic matter." What they mean is matter of the familiar sort—matter whose atoms have nuclei made from protons and neutrons.

Since baryonic matter is ordinary and all around us, it is natural to consider it first in discussing dark matter. Of the "usual suspects," it is the most usual, and there is no way we can absolutely rule out the possibility that dark matter is baryonic, although for reasons soon to be given this is not very likely.

Baryonic dark matter could take many forms. It could be a gas made mostly of hydrogen atoms and molecules. It could be a collection of Jupiter-sized objects orbiting out in the galactic halo. It could be brown dwarves (small stars that barely give off any radiation) or even black holes. Many such bodies could easily exist in the Milky Way without our having detected them. They would certainly constitute the least revolutionary answer to the question of the identity of dark matter.

The strongest evidence against a baryonic form of dark matter lies in the constraints placed on the universe by the formation of light nuclei three minutes into the Big Bang (see Chapter 3). You will recall that at this time the temperature was low enough so that if a proton and a neutron stuck together to form a nucleus, subsequent collisions would not be strong enough to tear them apart. The amount of light nuclei such as helium-4 and deuterium in the universe today served as one of the main pieces of evidence that substantiated the Big Bang view of the universe.

The rate of production of light nuclei at the three-minute mark depends on two things: the temperature and the density of matter. The temperature determines how fast the particles are moving when they collide, and hence whether newly formed nuclei will survive. It also tells us whether a proton will be moving fast enough when it approaches a nucleus to overcome the electrical repulsion between two positively charged particles. If it can, then the collision will result in a larger nucleus. If it can't, the proton will be pushed away and nothing will happen. The temperature is an index of what happens when a collision takes place.

The density of matter determines how often these collisions occur. If matter at the three-minute mark was tightly packed, collisions would be going on all the time. If particles were relatively sparse, on the other hand, collisions would be more rare. In the first case, more nuclei would have been produced; in the second, fewer. Thus, the density of particles that could make nuclei upon collision — what we have called baryonic matter — must have played an important role in determining how many light nuclei were in being after the universe was three minutes old.

In Chapter 3, we saw that the observed abundance of light nuclei was very close to what is predicted by the Big Bang theories. This constitutes evidence for the theory, of course, but it also shows that there is little leeway for changing things. If we increase the density by adding a lot of dark matter in the form of baryons, the rate of nucleus production will go up and the predicted values for helium abundance will rise beyond the limits placed by observation. If the universe is to be made of things like Jupiter-sized objects in galactic halos, the nuclei to make those objects must come from somewhere, and the period of nucleosynthesis at three minutes is the only source.

It turns out that the deuterium nucleus (one proton, one neutron) gives us a very good indication of matter densities in the early stages of the Big Bang. Deuterium is a heavier version of the ordinary hydrogen nucleus, and it is a component in what is called heavy water. For our purpose its most important property is that it isn't made in stars, although it can be burned up in stellar nuclear furnaces once it has been created in the Big Bang.* This means that if we find a deuterium nucleus, we can

---

* To be precise, deuterium is created in stars, but once a nucleus is made it collides almost immediately with another particle and is converted into a heavier nucleus.

be sure that it came to us direct from the early stages of the Big Bang. The deuterium need not be in stars, either; ocean water is just as good a source. This availability gives rise to what Nobel laureate Arno Penzias calls "doing astronomy with a shovel." You take a look at materials on the earth, see how much deuterium they contain, and extrapolate these numbers to the universe at large.

When this operation is carried out, we find that there is roughly one deuterium nucleus for every ten thousand normal hydrogen nuclei in the universe. This means that after the burst of nucleus building at three minutes was finished, there could not have been fewer deuterium nuclei than this amount. This, in turn, puts limits on the amount of baryonic matter in the universe. When the full calculation is done, we find that the observed abundance of deuterium requires the amount of baryonic matter in the universe to be less than 20 to 30 percent of the critical matter density defined in Chapter 8. Similar limits can be derived from looking at other nuclei. For technical reasons, it appears that the best (i.e., lowest) limits on the mass can be set by looking at lithium-7 abundances, which happen to be the hardest to measure and are therefore the most poorly known. The measurements of lithium-7 that we now have suggest that the limit may be closer to 20 percent than it is to 30 percent.

This result is extremely important. It is "tight" in the sense that we can look at the production of deuterium from the collisions of protons and neutrons in our laboratories, so there is little uncertainty about what happens as far as the nuclear-physics side of the problem goes. The only thing left unsure is the baryonic matter density, and for that we have one experimental number — the observed deuterium or lithium abundance — to tie things down. The result, therefore, is as hard as results ever get in cosmology.

We can say with some confidence, then, that the total amount of baryonic matter in the universe cannot exceed about 30 percent of the critical mass. You will recall that in Chapter 8 I showed this to be approximately equal to the amount of dark matter documented up to this point. It is thus barely possible that all of the dark matter is baryonic. I have to say, however, that I think it pretty unlikely that this will be the final view of things.

During the last two decades the amount of dark matter that can be documented has steadily increased, and at the same time, the limits imposed on baryonic mass from studies of light nuclei have steadily gone down. At the moment the two values are about the same, but it would take only one push from either side to get them past each other. As soon as the amount of dark matter exceeds the stated limits on baryonic matter, we are compelled to think of other forms the dark matter might take. Even if one wants to proceed with caution, simple prudence dictates that it is perhaps time to look beyond baryonic matter for a likelier "suspect."

Let us be a little more adventurous. Recall how the inflationary universe helped us establish, for the first time, a presumption in favor of flatness. There is now a theoretical as well as an aesthetic reason for believing that the universe has exactly the critical mass. If this is true, then at least 70 percent of that mass must be nonbaryonic. My own feeling is that it is highly likely the universe does possess the critical mass, and that most or all of the dark matter is nonbaryonic, but that's a personal opinion. You may choose to differ, and, given the frailty of the arguments on each side of the issue, I shan't argue.

A second reason for saying that dark matter is not baryonic was given in Chapter 4. As you recall, a universe that is made completely of this type of matter and in which large-scale structures exist will have a nonuniform cosmic microwave background. Since the background radiation in our universe is highly uniform, it follows that some of the dark matter must be nonbaryonic.

Finally, you can argue against baryonic dark matter by making what lawyers call a "merrie list." If galactic halos are baryonic, what forms can the matter take? It could be a gas, or it could be something held together chemically, like "snowballs" of frozen hydrogen or dust grains ranging from microscopic to planetary size. Finally, it could be some sort of body held together by gravity, such as Jupiter-sized objects ("planets" made primarily of hydrogen and helium) or dead stars ("cinders" which long ago burned out and quit emitting light). Another dark-matter candidate of long standing, the black hole, is formed at the end of a star's lifetime and can also be classified as a cinder.

If the galactic dark-matter halo were a gas, its temperature would have to be very high for the forces of pressure to overcome the inward pull of gravity. At these temperatures, the gas would emit X rays, and these would be easy to detect with modern satellite systems. Since the X rays are not seen, the halo cannot be a gas.

Frozen snowballs in space would not last: They would undergo a process called sublimation—a direct change from a solid to a gas. Sublimation is what occurs in dry ice all the time; the billowing clouds beloved of movie and stage directors are usually produced by allowing heaps of dry ice to become gaseous carbon dioxide. Sublimation of water ice is what dries clothes hung on a line when the temperature is below freezing. Because sublimation is possible, the halo cannot be snowballs. Nor can it be made of solids like dust grains or rocks, because if enough heavy material had been present to make such things when the galaxies formed, this same material would have had to have been incorporated in very old stars. These stars (called Population II by astronomers) are fourteen to eighteen billion years old and are very poor in heavy elements. How could there be heavy elements for grains and rocks, but none for the stars?

The Jupiter-sized objects are a little more difficult to eliminate. Jupiter was formed by a process similar to that which formed the sun—the condensation of a gas cloud under the influence of gravity. The obstacle to making the halo out of such objects is to explain how it was possible for billions of Jupiter-sized objects to be formed, but virtually no small stars. The point is that the planet Jupiter is very nearly a star. Had it accreted only a little more mass than it did, nuclear fires would have ignited in its interior and it would have become a star, albeit a small one. Why do we see no trace of faint stars slightly more massive than a large planet? What could be the cause of such a sharp cutoff in the size of objects in the halo? The lack of such a mechanism is a strong argument against this particular type of dark matter. It also serves as an argument against the brown-dwarf hypothesis, since brown dwarves are just on the other side of the star-planet borderline from Jupiter.

Finally, the stellar-cinder possibility can be discarded by noting that when a star dies, its death is almost always accompanied by the ejection of large amounts of material back into interstellar space. There is no evidence that such ejected matter is present in the halo, and this can be taken as proof that the halo is not made up of the corpses of old stars, either as burned-out cinders or as stellar black holes.

All these arguments — from nucleosynthesis, large-scale structure, and the elimination of various forms of baryonic matter — share a common fault: None completely shuts out the possibility that all the dark matter in the universe is baryonic. On the other hand, they do make it difficult to escape pitfalls in an all-baryonic universe. Ingenious theoreticians (and there are a good many of them around) have constructed complex schemes enabling each of the pitfalls just cited to be avoided. Such schemes carry a whiff of epicycles, the device used by medieval astronomers to adjust Ptolemy's universe to the observed facts. They may stave off disaster for a while, but they do not lead to any new science. Usually, when an idea is right, things fall into place as if by themselves — there's no need to contrive schemes for the disposal of embarrassing data.

For example, one could probably cook up a scheme in which all of the forms of baryonic dark matter listed above were mixed together to form the galactic halo. One could probably even adjust the proportions among the various types to avoid conflict with observations. But what would you have when you were finished? Nothing that anyone would want to pay attention to.

The reason is simple. As I pointed out in discussing the general theory of relativity, a successful theory in the sciences must be beautiful as well as workable. The kind of theory I imagined a moment ago doesn't meet this dual test. It may work, but it is profoundly ugly and unnatural. In the words of physicist Enrico Fermi, "It's not even wrong." It is the sort of thing that scientists would turn to only as a last resort, after everything else they could think of failed to work. It should come as no surprise, then, to learn that most theorists today have abandoned baryonic matter as the main constituent of the universe and have turned their attention elsewhere.

What About Neutrinos?

Once you rule out baryons, the next candidate for dark matter is the neutrino. I don't mean that the neutrino is familiar in the sense that we know about it from our everyday experience, but that it is a particle that physicists have known about for decades. The neutrino is produced in many nuclear reactions, including those that take place in the sun and those that were involved in the period of nucleosynthesis during the Big Bang.

For a rough idea of the number of neutrinos in the universe you can use a rule of thumb: One neutrino is around today for every nuclear reaction that ever took place. Calculations indicate that there were approximately a billion neutrinos produced during the Big Bang for every proton, neutron, or electron. Each volume of space the size of your body contains about ten million of these relic neutrinos, and that doesn't count the ones produced later in stars. Clearly, any particle as common as this could, in principle, have a major effect on the structure of the cosmos if it had a mass.

The existence of the neutrino was postulated in the 1930s. In the traditional theories, it is a particle that interacts very weakly with other kinds of matter. It has no mass, and therefore travels at the speed of light. More to the point for our purpose, its having zero mass means that it does not exert a gravitational force. Consequently, if the conventional theory is correct it makes little difference how many neutrinos may be lying around in the universe—they cannot affect things like gravitational collapse into galaxies or other structures.

In the early 1980s, however, some experimental results were announced that suggested that the neutrino, despite the strictures of the traditional theory, might have a tiny, nonzero mass. If this were the case, then multiplying even a small mass by the large number of neutrinos in the universe could well yield a total neutrino mass large enough to bring the mass density to its critical value.

The scenario for a neutrino-dominated universe went something like this: The neutrino is one of those particles whose interactions with matter become weaker as the temperature falls. This means that the neutrinos will stop exerting a force on (decouple from) ordinary matter long before ordinary radiation does. According to our current ideas, this happens about one second into the Big Bang. Afterward, the neutrinos expanded and cooled, thus creating a sort of mirror image of the cosmic microwave radiation.

One way to interpret the neutrino decoupling process is to say that it occurs when a neutrino traveling through matter is unlikely to interact with that matter. Two factors affect this likelihood: the density of the matter (which tells us how often neutrinos come near other particles) and the probability that a neutrino coming near another particle will actually interact. When the universe was one second old, the temperature was low enough so that the latter probability fell and the neutrinos formed an expanding, cooling cloud that no longer interacted with ordinary matter. If the neutrinos had a small mass, they could well clump together under the influence of gravity after this period, forming ready-made concentrations around which galaxies could later form. Thus it would seem that the massive neutrino, if it exists, is a perfect candidate for dark matter.

If neutrinos have a small mass, however, they will be moving at almost the speed of light when they decouple. This means that they will be the sort of dark matter we have labeled "hot." They will lead to a "top-down" theory of the universe, and all of the arguments advanced against hot dark matter in Chapter 7 could be invoked against them.

In addition, there are other, more detailed arguments that can be made against massive neutrinos as candidates for dark matter. These arguments, and others like them, have caused most cosmologists to turn away from massive neutrinos as a candidate for dark matter. They sound pretty convincing, but they pale in comparison to the most important reason to reject this hypothesis. Put as baldly as possible, there is no longer any unequivocal evidence that the neutrino has any mass other than zero.

It may seem strange to you that so much effort should have been expended on an idea that wasn't first verified in the laboratories, but the story of the massive neutrino is more complex than that. In fact, it is as good an illustration as I can find of the way that the mechanism of the scientific method works to eliminate incorrect results. So before going on, I would like to make a small digression and tell you about what I like to call "The Massive Neutrino Caper."

TEN

The Massive Neutrino Caper

Gild a farthing if you will,
Yet it is a farthing still.

—GILBERT AND SULLIVAN,
H.M.S. Pinafore

B<small>Y TRADITION</small>, the spring meeting of the American Physical
Society (the professional physicists' association) is in
Washington, D.C. By some impossible quirk of meteor-
ology, it always occurs one week after the cherry blossoms turn
the city into a setting for a picture postcard. The 1980 meeting
had a special significance for me because I took along my oldest
son, then a high school senior, to see the world of science. He
was college-bound and thinking seriously about becoming a
physicist, so it seemed a reasonable thing to have him attend.
Now I wonder whether what he saw there didn't influence his
decision to become an economist instead.

Physicists think of that particular convention as the "massive
neutrino meeting." The big event was the announcement by a
group of experimenters at the University of California at Irvine
that they had measurements indicating that the neutrino, a par-

ticle most physicists had thought to have zero mass, actually weighed something.* This news was exciting for two reasons: First, the man making the announcement was Frederick Reines, who had been the codiscoverer of the neutrino in 1956, and second, if the neutrino was really as massive as the group said, it might be the particle that closes the universe—the long-sought missing mass.

As the time for the announcement grew near, it became apparent that the rumors that filtered through the conference had borne results. It was impossible for everyone who wanted to hear Reines's paper to fit into the room assigned. There was a delay while the session was moved into a large ballroom. Since there wasn't time to get a slide projector, Reines was forced to spell out the names of the members of his group for the reporters present. He also had to describe his equations in words, writing them in the air with his finger as he did so. The experiment he described was a difficult affair, and the ways it could have failed were many. But if it was right, it meant that one of the generally accepted "truths" of modern physics was wrong.

To understand why so many physicists crowded into that ballroom on a fine spring day in Washington, you have to know a little about the neutrino. Its existence was first suggested back in the 1930s, when physicists were studying interactions in which something seemed to be amiss—there was sometimes more energy present before the interaction than after, for example. The neutrino ("little neutral one") was proposed as a way of patching things up. The hypothetical particle couldn't be detected itself, but its job was to carry away the excess energy and other things that seemed to be missing in the interaction. The situation was supposed to resemble a householder's situation when a burglar has stolen something from the house in the owner's absence—he knows someone has been there because something is missing, but nobody saw the thief.

It was twenty years before our detection systems became sensitive enough to find direct evidence that the neutrino existed (this was the 1956 experiment mentioned above). The neutrino

---

* The full story of the history and properties of the neutrino can be found in my *From Atoms to Quarks* (Scribners, 1980).

is indeed elusive; if one entered a bar of solid lead today, it could easily emerge four years from now at Alpha Centauri without having left a single disturbed atom to mark its passage. Despite this unwillingness to interact with matter, neutrinos can be routinely produced and measured in accelerator labs. With modern electronics they aren't nearly as hard to detect as they once were thought to be. Nevertheless, it is an important fact about the neutrino that its existence was postulated before it was actually found.

The neutrino has no electrical charge—if it had it would have been detected in the 1930s. It must be very light, otherwise some manifestation of its mass would be seen in the behavior of the particles it leaves behind. But a light particle—one with a small mass—is not the same as one with zero mass. Until very recently, if you had asked a physicist why the neutrino was believed to have zero mass, the answer would probably have been "Why not?" There was no evidence that the mass was anything other than zero, and zero is a nice round number, easy to remember and conjure with. If everything seems to work with a massless neutrino, why rock the boat?

But physicists are trained to be open-minded. If someone comes along with evidence that something generally believed is wrong, all other physicists are required to examine that evidence with care. This is especially true in cases like the supposed zero mass of the neutrino, where there's really no firm reason to hold the conventional belief.

Actually, there are different ways of finding out experimentally if the neutrino has mass or not. One is implied in the analogy of the burglar. The police can find out a lot about the burglar by carefully examining the scene of the crime. Just so, evidence for a nonzero mass for the neutrino can be turned up by carefully examining the particles that take part in reactions involving neutrinos. We shall take a look at some of these reactions later. A second method for fixing the neutrino mass involves something called "mixing." To understand this you need to know one more fact about neutrinos and one feature of quantum mechanics.

The fact is this: There is more than one type of neutrino. Physicists have seen two types in the laboratory, and theory tells us that there must be a third, rather rarer, species as well. These three neutrinos are distinguished from each other by the reactions in which they are formed and by the reactions which they initiate when they do interact (albeit not often) with matter. For example, in Fig. 10.1 we show the decay of a neutron. The neutron, left to itself, spontaneously transforms itself into a proton, an electron, and a neutrino. Because it is produced in association with an electron, it is called an electron neutrino.* If an electron neutrino encounters a proton, it can initiate a reaction in which a neutron and an anti-electron (positron) are created, as shown. This neutrino, in other words, is always associated with an electron, both when it is created and when it is destroyed.

The other kinds of neutrinos are associated with other leptons (see p. 45). We know that in addition to the electron (which is itself a lepton), there are two other types of leptons, called the mu and the tau mesons. We believe that there are three types of neutrinos in the universe, one associated with each of these leptons.

In such a situation, the laws of quantum mechanics suggest an interesting possibility. Consider first this analogy:

Suppose you have a stretch of three-lane highway which cars can enter through only one lane. If you were standing near the beginning of the highway, you would say that all the cars shared the property of being in the entrance lane—that there was only one kind of car present. As time went on, however, normal traffic processes would begin to move cars into the two empty lanes. One car would move out to pass, another would speed up and move into the outer lane, and so on. Eventually, the cars would be equally distributed in all three lanes. Someone standing a few miles from the entrance ramp would say that there were three kinds of cars present—one kind for each of the three lanes.

Under certain conditions, a beam of neutrinos could behave in the same way as the incoming line of cars. A collection of decaying neutrons could create a beam of pure electron neutri-

---

* The expert will realize that this particle in the figure is actually an antineutrino. The distinction is not important for our purposes.

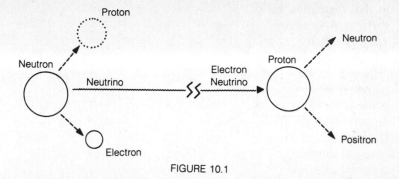

FIGURE 10.1

nos. As this beam progressed through space, the neutrinos could start to shift to other types, just as cars start to shift lanes. Eventually, you would have a beam made up of equal numbers of the three different types, just as eventually the cars fill all the lanes.

According to quantum mechanics, this sort of mixing among neutrino types can occur *only* if the masses of the neutrinos are different—that is, only if the electron neutrino has a different mass than the mu neutrino and the mu neutrino has a different mass than the tau neutrino. If, however, the conventional picture is correct and all three neutrinos have zero mass, then no mixing can happen—we would have a three-lane highway in which no cars could change lanes. Thus, evidence for mixing between neutrino types is also evidence for a nonzero mass for the neutrino. If mixing does occur, then at most one neutrino type can have zero mass. The other members of the mixed set must have masses other than zero.

This explains the governing idea of the experiment announced at the meeting in Washington. The setup is sketched in Fig. 10.2 (page 136). Neutrinos created by nuclear reactions in a research reactor stream out in all directions—nothing can stop them, for what's a few feet of concrete compared to a light-year of lead? Occasionally, one of the neutrinos initiates a reaction in a detector located (in this case) about thirty-six feet from the reactor core. The particles produced in this reaction are detected and the presence of the neutrino inferred.

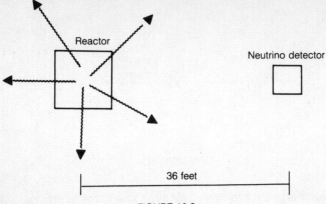

FIGURE 10.2

All the processes that go on in the nuclear reactor are like the neutron decay shown earlier in Fig. 10.1. They produce only electron neutrinos. Similarly, all the processes that go on in the detector are like the one shown on the right in Fig. 10.1 — they can be initiated only by electron neutrinos. Thus we have a system in which only electron neutrinos start out from the reactor, and only electron neutrinos are "seen" at the detector.

If the accepted view is right and the neutrinos are all massless, then no difficulties arise. Every electron neutrino produced in the reactor will still be an electron neutrino when it passes through the detector and has a chance to initiate reactions. But if some neutrinos have mass, then mixing will occur between the reactor and the detector. Figuratively speaking, some of the neutrinos will change lanes. This in turn means that the number of particles detected will fall, because although the total number of neutrinos streaming through the detector is unchanged, the number that initiate reactions (that is, the observed number of electron neutrinos) will drop. In our highway analogy, this is equivalent to the statement that the number of cars in the first lane will drop as the other two lanes fill up.

What the Irvine group found in their experiment was that the number of neutrinos arriving *and being detected* thirty-six feet from the reactor core was less than the number their calculations showed to be leaving the reactor. They interpreted this as evidence for the kind of mixing of neutrino types we've been discussing—the sort of thing that is called "neutrino oscillations" in the jargon of physics. Since oscillations can occur only if there is a mass difference between the different neutrino types, this was taken as evidence that neutrinos were not massless, as had previously been thought.

It was an extremely important suggestion. We know that early in the history of the Big Bang there were many nuclear reactions, and many of these produced neutrinos as a by-product, just as neutron decay does. Hence there are supposed to be many neutrinos out there—as many as 100 million for each normal massive particle. If each neutrino had a mass—even a tiny one—they could easily provide enough matter to close the universe. Thus at the very moment when conservative physicists were warning that the Irvine result had to be treated skeptically until confirmed by other laboratories, cosmologists were gearing up to explain the galaxy problem and the existence of dark matter by means of the newfound mass of the neutrino. Many of the arguments against the neutrino as the sole component of dark matter (see Chapter 9) were discovered during this burst of enthusiasm following the Washington meeting.

It is important to remember that although the discovery of oscillations in a neutrino beam may suggest that the neutrino has a mass, it doesn't tell you what that mass is. The speed with which neutrinos "shift lanes" in the beam is related to the *difference* in mass between the various neutrinos. So long as this difference is the same, the oscillations will be the same whether the neutrinos have a mass one-millionth that of an electron or the mass of an elephant.

This tells us why an announcement from the Institute for Experimental and Theoretical Physics in Moscow a few months after the Washington meeting played such an important role in the neutrino story. A group there had been working for over a decade to determine the neutrino mass by measuring very carefully what the neutrino removed from the reactions in which it figures—what I have called the burglar technique.

The scheme is this: When a neutron decays, as shown in Fig. 10.1, the energy is carried away by the three particles produced.

If you consider just the electron, it can have the complete range of energies — all the way from zero (when all the energy is carried off by the proton and neutrino) to the maximum energy possible in the process (when the electron carries off all the energy). The number of times you see an electron with a given energy, then, will be represented by a curve such as that to the left in Fig. 10.3. In most cases, the three particles will share the energy more or less equally; only occasionally will the electron carry away most of it. The section of the curve where this happens is boxed in the figure.

Now by looking closely at the boxed region we can learn something about the mass of the neutrino, because this mass is one of the things that determine the maximum energy the electron can have. The total energy available in the reaction is determined by the mass difference between the initial neutron and the masses of the final three particles. This mass difference, according to the equation $E = mc^2$, tells us how much energy the particles can carry off among them. If the neutrino has a mass, the amount of energy available to the other particles is reduced by the amount equivalent to that mass. Hence if the neutrino has a mass, the number of electrons in the boxed region should fall off more rapidly than if it doesn't. This is illustrated on the right in Fig. 10.3. By seeing to what extent the number of electrons gets smaller as the electron energy increases, you can not only see whether the neutrino has a nonzero mass, but actually determine what that mass is.

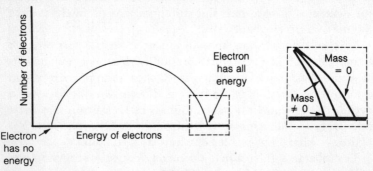

FIGURE 10.3

Before telling you what the Russian group announced, I must speak of the units in which the neutrino mass is measured. The electron volt (eV) is the amount of energy it takes to move one electron through one volt. Moving an electron from one pole of your car battery to the other, for example, requires 12 eV of energy. Since mass and energy are equivalent, it is possible to measure the masses of particles in eV through a simple application of $E = mc^2$. The mass of the electron, for example, is about 550,000 eV, while the mass of the proton is 939 million eV. The masses of the neutrinos under discussion are all in the range of 0 to 50 eV — 100,000 times lighter than the electron itself.

What the Russians announced was this: According to their data, the mass of the neutrino was less than 46 eV, but greater than 14 eV. The significant point was that the mass could *not* be zero. Coupled with the Irvine evidence for oscillations, the Russian announcement fueled speculation that neutrinos might be the main component of dark matter.

But physicists are a cantankerous lot. A new finding such as the nonzero neutrino mass is often looked upon as a target to shoot at instead of as a scientific advance. The first thing experimentalists are likely to ask is "What could have fooled you into thinking you had a result when you really hadn't?" No group of scientists, I am convinced, takes this question more seriously than experimental physicists; none work so hard to disprove their own results; none subject the results of others to such sustained attack. This isn't done in a personal way; indeed, most physicists get along together as well as any other group of professionals. The conflict is simply seen as the most efficient way of advancing the science. Nothing can be accepted until it has been subjected to exhaustive collective criticism. You can't understand the next step in the massive neutrino story unless you keep this characteristic of the physics community in mind.

The two experiments announced in 1980 and 1981 looked impressive, especially since they came to the same conclusion through two different methods. Each, to be sure, had inherent uncertainties. In the Irvine experiment, for example, the conclusion rested on the statement that fewer neutrinos were detected than one would expect. What if an error had been made in calculating the number of neutrinos leaving the reactor? Furthermore, neutrinos aren't the only particles that come from a reactor core — some neutrons would be mixed in, contaminating the beam. This sort of problem is routinely dealt with by using electronic systems to pick out the particles you want, but when dealing with something as elusive as the neutrino, you worry about the likelihood that you have systematically included or excluded some wrong events in this procedure. The only way to settle these doubts is to repeat the experiment (preferably at a different reactor) in a situation in which the detector can be moved, so that the beam can be measured first in one spot, then in another spot farther downstream. In this way, any errors of the type we've mentioned would tend to cancel out, and the oscillation would be measured between the two positions with high accuracy.

That the Irvine group had not done the experiment in this fashion was not considered a criticism. They had carried out their work according to accepted procedures, and they themselves pointed out the need for the two-position experiment. But to move a multiton neutrino detector around in the crowded research bay of a major reactor facility is no small task. Indeed, some groups who now do this sort of work find it easier to build two huge detectors than to try to move one. In any event, the Irvine result could not be accepted until it had been verified in just this way.

The Moscow measurement also had inherent problems. I have spoken rather loosely of measuring the decay of a neutron; the fact is, there are no large collections of free neutrons in nature. In practice, the Russian experiment was done by measuring the decay of tritium, an atom whose nucleus has one proton and two neutrons. Tritium is an isotope of hydrogen and hence can be bound chemically into any substance in which hydrogen is normally found. The tritium that was measured was bound into a complex organic molecule called valine.

But this scheme introduces a new kind of problem. When a neutron decays inside a tritium nucleus within a large molecule, energy can be transferred to the molecule as well as to the particles coming out. After the decay, for example, the molecule might be vibrating like a plucked string. This transfer of energy would lower the maximum energy that the electron could have, and this shortfall could masquerade as a neutrino mass. The only way to correct for this possible confusion (other than by doing theoretical calculations) would be to repeat the experiment with the tritium in a different kind of environment—either in a different substance or by itself.

By the beginning of 1982, then, it was clear that despite the positive results of these two experiments, a lot of work would have to be done before a massive neutrino was accepted. You may be wondering why, in the light of all the uncertainties, the theoretical community didn't adopt a wait-and-see attitude before starting to work on universes in which massive neutrinos played the role of dark matter. The answer has to do with the way the theoretical physics (and astrophysics) communities work.

Theory is a young man's game. There are notable exceptions, of course, but by and large the career of a theoretical physicist is determined by what is accomplished the first decade after graduate school—say between the ages of 25 and 35. In this short span, the aspiring physicist must do something spectacular to bring himself to the attention of his colleagues around the world, and one of the best ways to do this is to be the first to exploit a new experimental finding. If you write one of the first papers detailing how the galaxy problem can be solved with massive neutrinos, you gain at once the marks of academic success—invitations to visit other universities to explain your work, large research grants from funding agencies, and, if you are very lucky, an invitation to summarize the work in the new field for *Scientific American*. But to get all these rewards, you have to be in the field quickly with a major piece of research. You have to be seen as the pioneer.

This accounts for the tremendous pressure on theoretical physicists—both the young ones who want to establish a reputation and the older ones who want to maintain their standing—to move in quickly when a new field opens up. And hence the massive neutrino caper—theorists spinning out whole new universes, one after the other, on the basis of some very preliminary experimental results. Given the way the community operates, this behavior makes good sense: it's a rational way of looking at the odds. If you work out the consequences of a result that later turns out to be wrong, you lose nothing. The papers you publish will attract some momentary interest, and no one will blame you for basing them on a false experimental finding. After all, the incorrect experiment wasn't *your* fault. If, on the other hand, the result is right, you stand to win in a big way. The one sure thing is that if you wait until the experimental result is well established, you have definitely lost, for others will have taken the gamble and made the major discoveries. All that will be left to do is adding footnotes to their work.

This way of working guarantees that even a rumor of a new experimental finding is likely to start a stampede of researchers into a new corner of the field. It is this that suggests faddishness in science, and it explains why ideas that the media tout as final one year disappear in the next, never to be heard of again. Such was to be the fate of the massive neutrino.

While cosmologists were busily working out the consequences of neutrinos with masses in the neighborhood of 30 eV, research groups at reactors around the world were no less busily checking the Irvine result. The first fresh announcements came from "quick and dirty" tests: experimenters halted ongoing projects and quickly adapted their equipment to take a look at oscillations. During this early period, the situation was a little murky. Some people seemed to support the idea of oscillations, some didn't. But as time went on, experiments designed specifically for the task went on line and the tide began to turn. Slowly the limits on the presence of neutrino oscillations in reactor beams were pushed down. It became clear that the original results were simply wrong (although I haven't heard anyone explain why), and that if neutrino oscillations do indeed exist in nature they are not going to be seen by any but the most delicate and difficult experiment. By 1984, in the same room where the first positive result had been announced, Felix Boehm of Caltech gave a review talk, summarizing the experimental situation and laying neutrino oscillations firmly to rest.

This put the cosmologists in an uncomfortable position. Even though the oscillation experiment didn't give a value for the neutrino mass, it lent credibility to the Russian result, which did give such a value. Now speculations on the possible role of massive neutrinos in the universe had to be based on that result alone—one experiment. Furthermore, although it is usually considered impolite to say so in public, Western scientists have no great confidence in experimental work done in the Soviet Union, especially when that work involves delicate measurements and requires sophisticated electronic equipment. The Soviets don't have a good track record in this sort of thing, probably because, unlike their colleagues in the West, they are compelled to work with whatever electronic equipment isn't needed by their military.

During the mid 1980s, interest in massive neutrinos slowly died out among cosmologists. Theorists looking for the big strike moved over into cold dark matter (see Chapter 7) and then on to cosmic strings (Chapter 12). Occasional lip service was paid to the Russian result, but few people took the massive neutrino seriously anymore. At the same time, several laboratories around the world were going through the laborious process of setting up equipment to repeat the measurements of the electrons coming from the decay of neutrons in tritium. In Zurich, the tritium was built into a special carbon-based material, but not locked into a molecule as it was in the Russian experiment. At Los Alamos, researchers made their measurements on tritium gas. The Zurich technique had the advantage of being able to measure lots of tritium decays, but had to worry about the effects of the carbon material. The Los Alamos group had clean tritium, but had to be content with seeing fewer decays because the tritium was in a gas instead of a solid.

In the summer of 1986, the results from these experiments were announced. While the Zurich and Los Alamos groups continue to argue about whose procedure is best, they agree on two things: first, that the possibility of the neutrino mass being zero (or very small) is consistent with the data, and second, that the mass must in any case be less than about 18 eV (28 eV for the Los Alamos experiment).

In other words, the second wave of experiments contradicts the Russian result and its claim that the neutrino mass must be *bigger* than 14 eV. Most physicists expect that the limits on the neutrino mass will be pushed down still farther as additional experiments are done. When these limits get below a few eV, they cease to beome "cosmologically interesting"—that is, if the masses are that low, the neutrinos would not be able to close the universe. I suspect that if a poll were taken among physicists today, the majority would see this conclusion as the most likely outcome of the massive neutrino caper.

What may we conclude from all of this? The surge of interest that began on that beautiful spring afternoon in 1980 has spent itself. A decade hence, no one but historians will remember that for a short time cosmologists flirted with the idea of massive neutrinos as a major component of dark matter. That ephemeral supposition has contributed little to our understanding of the universe, but it should contribute a great deal to our understanding of the way science works.

Depending on your temperament, you will "view with alarm" or "point with pride." If the former, you will think of all the wasted effort and wheel-spinning that went on while the experimental findings were being sorted out and all the talent expended for naught. If the latter is your view, you will rejoice that in a few short years a very complex situation was clarified by the concerted efforts of the international physics community. The mass of the neutrino was shown to be small and consistent with zero, and the ways of gaining further knowledge about this fundamental constant of nature were laid out. You would be entitled to argue that a few years' confusion was a small price to pay for this outcome.

Whatever position you take, it is clear that no matter what the solution to the dark-matter problem is, it will not involve only massive neutrinos, and it may not involve neutrinos at all. We must move on to look for other, more exotic possibilities.

ELEVEN

Is the Universe Controlled by Wimps? Exotic Candidates for Dark Matter

The unicorn is a mythical beast.
—JAMES THURBER,
The Unicorn in the Garden

THE PRESENT STATE of affairs is that we know there is a great deal of dark matter in the universe, but we've managed to rule out every ordinary type of candidate particle that we know of and can produce in our laboratories. Under such circumstances, we can come to no other conclusion than that the dark matter must exist in some form which we have not yet seen and whose properties we are totally ignorant of. It might entertain you if I reported that, naturally, this predicament has caused great consternation among scholars, that in solemn conclave assembled they shake their heads and lament the failure of our search for the ultimate components of the universe. I'd like to report this and fulfill your expectation, but I can't, for the truth is that cosmologists, finding themselves in a blind alley, have behaved the way a kid does in front of a pile of new toys at Christmas. Theorists like nothing better than a situation in which they can let their imagination run wild without fear that anything as rude as an experiment or observation will end their game. In any event, they have produced extraordinary suggestions as to what the dark matter of the universe may be.

The way they go about it is this: They take a currently fashionable theory of the interactions of the fundamental constituents of matter and note that the theory either requires or allows the existence of some sort of new particle. The requirements for the nature of this as yet undiscovered particle are examined, and if it could possibly play the role of cold dark matter as outlined in Chapter 7, it is announced with great fanfare that the ultimate constituent of the universe has been discovered.

This mode of search illustrates as well as anything the coming together of particle physics and cosmology. The existence of each particle we're about to discuss was originally suggested for reasons that have nothing to do with the structure of the universe. Work on their properties was propelled solely by the internal requirements of the theories being forged to explain the interactions between fundamental particles. Only after these steps had been completed was it realized that these particles could also play a cosmological role.

Here, then, are a few of the particles proposed as constituting dark matter. They are referred to collectively as WIMPs, an acronym for Weakly Interacting Massive Particle. When you have looked over the reputable candidates suggested over the past few years, I will leave it to you to guess what the disreputable ones are like. But before beginning this exercise in imagination, I want to make one point emphatically: None of the forms of matter to be mentioned—not a single one—has ever been seen in a laboratory. You may think they ought to exist, you may even argue that if we just searched hard enough we would find them; but some medieval scientists thought the same about the unicorn.

Since the description of more exotic possibilities will lead us into some rather abstract byways of elementary-particle physics, the reader who wishes to be spared this adventure can skip ahead to the summary on page 155, where I have listed the candidates for dark matter and their properties.

Supersymmetry

By far the greatest number of candidates for dark matter arise from a principle known as supersymmetry. The theories that presuppose supersymmetry are those that unify all four forces — the ultimate theories that govern the first instant of the life of the universe (see page 46). In the lighthearted jargon of modern cosmology, they are sometimes referred to as the TOE, for the Theory of Everything.

What is supersymmetry? When matter is broken down into its ultimate constituents, we recognize two kinds of elementary particles. First, there are the quarks and particles like the electron (leptons) that make up solid matter. These particles are grouped under the general term "fermions," after Enrico Fermi, the Italian-American physicist who first investigated their properties. They are characterized by the fact that they spin around their axes of rotation at rates which are half-integer fractions of a basic unit of rotation. In other words, they have spin ½, ³⁄₂, ⁵⁄₂, and so on, but never 1, 2, 3 . . .

The second class of particles are called "bosons," after the Indian physicist S. N. Bose. These particles have spin 0, 1, 2, and so on. Unlike the fermions, they are not part of the structure of solid matter. Instead, they flit between other particles, creating the forces that bind matter together (or, on occasion, tear it apart). The most familiar boson is the photon, the particle associated with ordinary light. When photons are exchanged back and forth between two charged particles (for example, an electron and the nucleus around which it orbits) the exchange creates the familiar electrical force. The atom can thus be thought of as held together by the photons being exchanged between the electrons and the nucleus.

In the atom we see the roles of the two types of particles most clearly. The structure of the atom — the solid stuff of which it is made — is composed of electrons, protons, and neutrons. All these are fermions. So are the quarks that make up the protons and neutrons. But these particles are held together in their structure by the constant exchange of bosons. Just as photons keep the electrons in orbit, analogous particles called gluons keep the particles in the nucleus together.*

The point about bosons and fermions is this: We never see an interaction in the laboratory in which one sort of particle changes into the other. It appears, in other words, that an impenetrable wall stands between the two classes of particles — they are forever divided according to the function they perform. This distinction must have held since gravity froze away from the other forces, a time when the universe was 10^{-43} second old.

For various technical reasons, it turns out that if we want to write down the ultimate unified theory, a theory in which gravity is treated in the same way as all the other forces, we must introduce reactions in which fermions can turn into bosons and bosons can turn into fermions. In effect, the distinction between particles-as-structure and particles-as-force must not have been present when the universe was born, and must have appeared after the first freezing, when gravity separated from all the other forces. (We should note that when particles can be converted into one another in any sort of interaction, physicists think of them as being the same particle. In a similar way, you are the same person whether dressed in a business suit or sweat clothes.)

A world in which the distinction between bosons and fermions does not hold is said to be supersymmetric. It would be a world of ultimate simplicity, because there would be only one sort of particle, and it would account for both structure and force. The most promising approach to understanding the origins of the universe seem to involve theories which postulate that everything began in a supersymmetric state.

Such theories also predict that in the beginning, when the universe was supersymmetric, there were partners — mirror images, so to speak—of all the familiar particles. We know that in

---

* A more complete description of the way the exchange of particles can give rise to a force is given in my book *From Atoms to Quarks* (Scribners, 1980).

our world today there is a boson called the photon which generates the electrical force. The supersymmetry theories say that before gravity froze away from the other forces, there was another particle, identical to the photon in every way, except that it had spin ½ instead of spin 1. This other particle, called the photino, was a fermion. In the early universe, this particle and the photon could be transformed into one another.

When gravity froze away from the other forces, the symmetry between bosons and fermions disappeared and the early simplicity of the universe was lost. From the point of view of the particles, this loss of symmetry manifested itself in a process by which the photino became very massive—much heavier than the proton. The theories predict that in today's universe there is a kind of mirror world made up of the supersymmetric partners of all the particles we normally see. We know that there is a particle called the electron, for example, but the theories tell us that it should also be possible to create a supersymmetric analogue of the electron which has spin 1 instead of spin ½ and is very massive. This particle is called the selectron. There are also supposed to be squarks (the analogues of quarks), sneutrinos (the analogues of neutrinos), and so on. Perhaps there are even smen and swomen, although the theories have not, so far as I know, addressed that question yet.

Now, the theories do not require that these supersymmetric particles congregate in the same places as ordinary matter. Neither do they give us any firm notion as to how much mass something like the photino is supposed to have, although current thinking is that the photino is probably at least forty times as massive as the proton. At the same time, the theories require that once the symmetry is broken the interaction between the supersymmetric world and our own must be very weak. All the "sparticles" should be much more elusive than the neutrino, and impossible to detect directly with our present technology.

With all these features, the supersymmetric particles are perfect candidates for dark matter. They are massive, so they can exert a gravitational force. They interact weakly, so they would not interfere with the normal workings of the things like stars or high energy accelerators. What more could one ask?

Superstrings

The currently fashionable realization of the idea of supersymmetry is contained in what are called "superstring" theories. In these theories the basic constituents of all particles are tiny strings of very dense matter buried down inside a fluffy cloud of stuff that makes up the outer layers of the familiar particles. The strings are very small—they are no more than 10^{-33} centimeters long. The kind of string that is supposed to be the heart of matter thus bears about the same size relation to a proton as you bear to a small galaxy. In the first string theories formulated, fermions corresponded to loops such as that shown on the left in Fig. 11.1, while bosons corresponded to open strings such as that shown on the right.* Even quarks are supposed to be made from strings if you look at them closely enough.

If we believe that the heart of matter has some sort of stringy structure, then there is a useful analogy that helps us understand how matter (particularly supersymmetric matter) must behave. When you pluck a guitar string, you can make it vibrate as shown on the bottom right drawing in Fig. 11.2. This is called the fundamental mode and is the lowest-frequency note the string can produce. You can also make the string vibrate in other configurations, as shown in the other three drawings in the figure. These configurations produce the overtones and harmonics that give the note its richness and a particular instrument its timbre.

It requires energy to set a guitar string in motion, and this energy manifests itself in the energy of motion of the string as it vibrates. Since energy and mass are equivalent ($E=mc^2$), this means that each of the modes of vibration of the guitar string shown has a slightly different mass associated with it.

When a superstring is "plucked," it, too, can vibrate in many different modes. As with the guitar string, each of these modes will have a different energy and mass than the others. When we look at a vibrating superstring, then, we see something that has

* In more modern versions, all the strings are loops and the fermions and bosons correspond to waves traveling in clockwise and counterclockwise directions, respectively.

FIGURE 11.1

mass, and that mass differs from one vibrating string to another. But this is precisely what we see when we look at different particles — they, too, have different masses. This explains one of the central tenets of the superstring theory. Each of the infinite possible modes of vibration of the string will correspond to a different particle, so we expect that there is an infinite number of possible particles in the world.

This notion also leads us to suspect that if a string is plucked (for example, by having energy added in a high-energy collision), the higher harmonics will eventually die out, leaving only the fundamental mode of vibration. This is an important point, because it means that if we look for supersymmetric particles, we are likely to find only those corresponding to the fundamental mode of oscillation, which we interpret as being the lowest-mass supersymmetric particle.

So far, this discussion of superstrings may sound suspiciously

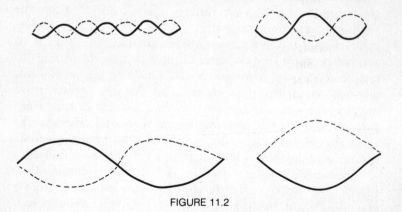

FIGURE 11.2

fanciful, but there is nothing really strange about it. Now let me drop the other shoe. In all the superstring theories being considered by theorists today, the strings do not vibrate in the ordinary three dimensions, nor even in the physicist's world of four. The theories seem to be telling us that the strings have to vibrate in either ten or twenty-six dimensions. (Are you *sure* you don't want to skip ahead to the summary on page 155?)

Don't even try to visualize more than three dimensions—it can't be done. The theorists are driven to considering these sorts of things because it is only in higher dimensions that the theories they write down avoid having something called anomalies. Never mind the technical definition of this term: A mathematician reacts to an anomaly in his equations the way you react to notification that you've overdrawn your checking account. Anomalies are bad things and are to be avoided at all costs, even if it means piling on the dimensions.

Without making a fuss about multidimensionality, I should point out that there is no reason why we should expect to get rid of anomalies in *any* dimension. What the theorists found is similar to what you would find if you discovered that you could balance your checkbook if you used paper with ten or twenty-six lines on it, but under no other circumstances. You might eventually understand why your checking account required this, but you would always feel a sense of mystery in the entire business. Likewise, theorists now understand why ten and twenty-six dimensional strings are different, but it doesn't make the result any less astonishing.

Of course, the multidimensionality of the strings leads to another problem. We do, after all, live in a world of four dimensions (three in space and one of time). To get around this disparity, superstring theorists postulate that when gravity froze out, the extra dimensions underwent a process called "compactification." As the name implies, the theories predict that the extra dimensions "curled up," so that the world appears four-dimensional unless you look at it on a very fine scale indeed.

The standard analogy used to explain compactification involves an ordinary garden hose. When you see a hose from a great distance, it looks like a string—a one-dimensional object.

When you get up to it, you see that there is an extra dimension —the width of the hose—that is "curled up" and visible only on close inspection. So does a normal particle appear to be four-dimensional unless you look at it closely enough to make out the string at its heart, in which case it appears to be ten-dimensional.

What are we to make of superstrings? On the one hand, they provide a truly beautiful and elegant Theory of Everything. They offer a scheme in which all the forces appear on an equal footing, the ultimate realization of Einstein's dream. They even have the advantage that in some versions gravity cannot be neglected even after the Planck time, the time when gravity unifies with all the other forces so that they describe forces in a unified way even in our present world.

On the other hand, there is at present no theoretical prediction that can be tested by experiment or observation. The theorists have gotten so far out in front of the experimenters that they can no longer seek guidance from observation, but must rely on their aesthetic sense. This way of doing science has never been tried before, and it will be interesting to see what comes of it.

The Shadow Universe

One striking possible outcome of superstring theories is that they may give rise to yet another kind of dark matter. Rocky Kolb, David Seckel, and Michael Turner at the University of Chicago and the Fermi National Accelerator Laboratory pointed out that in one version of the superstring theory, which is particularly appealing from an aesthetic point of view, the equations seem to suggest that at the Planck time the universe split into two separate parts. There is our normal world, with its full complement of particles and supersymmetric partners, and there is, in addition, a shadow world. The matter in this shadow world bears a resemblance to ours in that it, too, has its particles and "sparticles." Within each world, particles interact with each other through a full complement of four forces. Particles in one world, however, can interact with particles in the other only through the force of gravity.* An electron and a shadow electron can be near each other and not feel an electrical force, even though each is carrying its own version of electrical charge. The only force between the two would be the relatively weak force of gravity.

The idea of a shadow universe gives us a simple way of thinking about dark matter. The universe split into matter and shadow matter at the Planck time, and each evolved according to its own laws. Presumably some shadow Hubble discovered that his shadow universe was expanding, and presumably some shadow astronomers think of us as candidates for their dark matter. Maybe there is even a shadow you out there somewhere reading a shadow version of this book!

Axions — Another Dark Horse

Another favorite WIMP is called the axion. Like the photino and its partners, the axion was suggested by considerations of symmetry. Unlike sparticles, however, it comes out of the Grand Unified Theories that describe the universe at 10^{-35} second, rather than out of the fully unified theories that operate at the Planck time.

It has long been known to physicists that every reaction between elementary particles obeys a symmetry we call CPT. This means that if we look at the film of a reaction, and then look at the same reaction when we (1) view it in a mirror, (2) replace all particles by antiparticles, and (3) run the film backward, the results will be identical. In this scheme the P stands for parity (the mirror), C stands for charge conjugation (putting in the antiparticles), and T for time reversal (running the film backward).

It used to be thought that the world was symmetric under CPT because, at least at the level of elementary particles, it was

---

* Because of this property, shadow matter should not, strictly speaking, be included under the heading of WIMP, since WIMPs can interact with normal matter through forces other than gravity. Shadow matter would still, however, be "dark," in the sense in which we are using that term.

symmetric under C, P, and T independently. It has turned out that this is not the case. The world seen in a mirror deviates slightly from the world viewed straight on, as does the world seen when the film is run backward. What happens is that the deviations between the real and reversed world in each of these cases cancel each other out when we look at the three reversals combined.

Although this is true, it is also true that the world is *almost* symmetric under CP acting alone and T acting alone. That is, the world is almost the same if you look in a mirror and replace particles with antiparticles as it is if you look straight on. It's the "almost" that bothers physicists. Why should things be so close to perfect, but just miss?

In 1977, Roberto Peccei and Helen Quinn, both at Stanford University, found a natural way to answer this question in terms of the Grand Unified Theories, a way that was later realized to involve the existence of a new and as yet undiscovered particle. The particle they suggested has been dubbed the "axion." The axion is supposed to be very light (less than a millionth the mass of the electron) and to interact only very weakly with other matter. It is the small mass and weak interaction that explain the "almost" that was troubling theorists.

Calculations by cosmologists show that in an expanding universe, axions would be expected to form a background radiation field something like the 3-degree microwave background (see Chapter 3). It is irregularities in this axion background that could play the role of dark matter.

Summary

Below I list the candidates for dark matter, together with a brief description of their properties and a short statement explaining why they are thought to exist.

SUPERSYMMETRIC PARTICLES—PHOTINOS, SQUARKS, ETC.

These particles are predicted by theories that unify all the forces of nature. They form a set of counterparts of the particles with which we are familiar, but are much heavier. They are named in analogy to their partners: the squark is the supersymmetric partner of the quark, the photino the partner of the photon, and so on. The lightest of these particles could be the dark matter. If so, each particle probably weighs at least forty times as much as the proton.

SHADOW MATTER

In some versions of the so-called superstring theories, there is a whole universe of shadow matter existing in parallel with our own. The two universes separated when gravity froze out from the other forces. The shadow particles interact with us only through the force of gravity, which makes them ideal candidates for dark matter.

AXIONS

The axion is a very light (but presumably very common) particle which, if it existed, would solve a long-standing problem in the theory of elementary particles. It is estimated to have a mass less than a millionth that of the electron, and is supposed to pervade the universe in a manner similar to the microwave background. Dark matter would consist of aggregations of axions above the general background level.

WIMPs in the Sun?

Throughout this chapter, I have emphasized the fact that all the candidates for dark matter that we have discussed are purely hypothetical particles. There is no evidence that any of them actually are to be found in nature. I would be remiss, however, if I didn't mention one argument—one tiny glimmer of hope —that tends to support the existence of WIMPs of one sort or another. This argument has to do with some problems that have arisen in our understanding of the workings and structure of the sun.

We believe that the sun's energy comes from nuclear reactions deep within its core. If this is indeed the case, then theory tells us that those reactions should be producing neutrinos—neutrinos that are in principle detectable on the earth. If we know the temperature and composition of the core (as we believe we do), then we can predict exactly how many neutrinos we should detect. For the last twenty years an experiment has been running in a gold mine in South Dakota detecting these very neutrinos, and unfortunately, the results are puzzling. The detected number is only about a third of what was expected. This is known as the solar neutrino problem.

The second characteristic of the sun that bears on the existence of WIMPs goes by the name of solar oscillations. When astronomers watch the solar surface carefully, they see it vibrate and jiggle—the whole sun can pulsate over periods of several hours. These oscillations are the analogue of earthquake waves, and astronomers call their studies "solar seismology." Since we believe we know the composition of the sun, we ought to be able to predict the properties of these "sunquake" waves. There are, however, some long-standing discrepancies between theory and observation in this field.

Recently, astronomers have noted that if the galaxy is really full of dark matter in the form of WIMPs, then during its lifetime the sun would have absorbed a fair number of them. The WIMPs would therefore be part of the sun's composition—a part that hasn't been taken into account up to now. When WIMPs are included in calculations, it turns out that two things follow: First, the temperature at the core of the sun turns out to be less than we thought, so that fewer neutrinos are emitted, and second, the properties of the bulk of the sun are changed in such a way as to make the predictions of the solar oscillations come out right.

This result is a straw in the wind as far as the existence of WIMPs is concerned, but don't give it too much weight. Both the neutrino problem and the oscillations could be explained equally well by other effects that have nothing to do with WIMPs. For example, the kind of neutrino oscillations discussed in Chapter 10 could solve the solar neutrino problem if the neutrino had even a very small mass, and various changes in the details of the internal structure of the sun could explain the oscillations.

Nevertheless, these solar phenomena constitute the only indication we have that one of the exotic candidates for dark matter may actually exist.

Dinosaurs and Dark Matter

All this talk of supersymmetry and ultimate theories gives the discussion of the nature of dark matter a ponderous tone that bears no resemblance to the way the debate actually proceeds among cosmologists. One of the things I like most about the field is that everyone in it seems to be able to keep a sense of humor and perspective about his or her work.

A while ago, talking to a group of cosmologists about the extinction of the dinosaurs, I explained that one theory held that the sun, in its rotation around the Milky Way, periodically moved up out of the plane of the galaxy. When it did so, the dust that exists in the plane might cease to protect the earth, which would then be bathed in lethal cosmic rays that the authors of the theory thought might permeate the cosmos. From the back of the room, a graduate student popped up with "You mean the dinosaurs were wiped out by photino radiation?"

We all burst out laughing. The juxtaposition of the large, somber, and indubitably real fossils in museums with the airy, theoretical photino was ludicrous. This playful approach to one of the major questions in modern cosmology is, I am happy to say, typical of the state of mind of the scientists who work in it. It helps make a very complex subject highly enjoyable.

TWELVE

Cosmic Strings
Solution or Snake Oil?

Hush, boys, and belt your gobs,
I'll tell yez all an awful story.
Hush, boys, and belt your gobs,
I'll tell yez 'bout the worm.

—Irish folk song

W E SHOULD GET ONE THING clear right from the start. Cosmic strings, the subject of this chapter, and the superstrings I talked about in Chapter 11 are two very different things. The only connection between them is their names. Superstrings are smaller than the smallest elementary particle, but cosmic strings may span large parts of the universe. In some versions of the theories, in fact, they run through the entire universe like a string through a pearl necklace. In fanciful moments, I like to think of them as a reincarnation of the mythical Worm Ourobouros. This was an ancient Egyptian symbol consisting of a snake eating its own tail. In Norse cosmology, the worm surrounded the world, forever unseen but forever exerting its influence on matters terrestrial. Anyone willing to stretch a few points will be able to see the connection between the old myths and the new constructs in cosmology.

159

What Are Strings?

As the name implies, cosmic strings are long, one-dimensional objects in space. If they exist (and, as we shall see, that's a big if), they are incredibly massive. At the surface of the earth a piece of cosmic string long enough to stretch from one side of an atom to the other would weigh a million tons. A piece the size of a grain of sand would require a line of dump trucks wrapped eight times around the equator to support it. Because of its incredible mass, the string exerts a strong gravitational attraction on material around it. Consequently, it is ideally suited to play the role of dark matter in the formation of large-scale structure in the universe.

The large mass of the string tells us that it must have been created very early in the life of the universe, when temperatures were high and there was plenty of energy available to make exotic objects. Certainly, no process in our present universe, be it man-made or natural, could produce the energies needed to make a cosmic string. If such objects exist, they have to be remnants of an earlier time.

Not only are strings more massive than anything we can create, but if they exist they are matter in a totally new form. The usual picture of a particle such as a proton is of a small blob of "stuff." We know intellectually that the proton can equally well be considered a bundle of pure energy — $E = mc^2$ — but it is very hard to imagine it that way. In most situations, this mass-energy equivalence causes no difficulties. With the cosmic string, however, we have to confront it at a fundamental level, which is anything but easy.

The Formation of Cosmic Strings

To understand strings we must go back to the time 10^{-35} seconds after the beginning, when the strong force froze out and the universe inflated. Strings may be regarded as a by-product of the freezing process itself.

When water freezes, it goes from a state of high symmetry to one of lower symmetry. What I mean is this: If you were in the middle of an ice crystal you could turn the crystal 60 degrees and your surroundings would be identical in appearance with what they were to start with. This is what we mean when we say that a snowflake is symmetrical. But this is a limited kind of symmetry—if you turned the crystal 10 degrees or 34 degrees, you would see immediately that your surroundings had changed. On the other hand, if you were suspended in water this limitation would not be there. No matter which way you turned, you would see exactly the same thing. To a physicist, therefore, a drop of water, bland and uniform though it may be, has a higher symmetry than a snowflake or ice crystal. I am belaboring the point because the physicist's way of using the term "symmetric" is not that of everyday speech.

The freezing of water, then, can be thought of as a transition from higher to lower symmetry. Similarly, the GUT freezing can be thought of as a transition from a universe of higher symmetry (one in which there were only two forces in nature) to a universe of lower symmetry (in which there were three). As we saw in Chapter 11, modern theories of supersymmetry rely explicitly on this sort of thinking. We can learn something about the GUT freezing, then, by watching water freeze on an open pond.

The surface of the pond does not suddenly turn into a sheet of ice when the temperature drops below 32 degrees. Instead, the ice grows outward from various spots on the shore where the water, being shallow, cools most rapidly. Within any patch of ice, the crystals are lined up in the same direction, but the crystals in one patch need not be in the same direction as those in another. It would be remarkable if they were, because it would mean that a patch of ice growing on one side of the pond knew what a patch on the other side was doing. Otherwise, how could it line itself up the right way?

As the temperature keeps falling, the patches grow until they suture themselves together and cover the entire pond. It is interesting to ponder what happens when one patch meets another. If we mark the direction of one of the axes of symmetry of the crystals in each patch by an arrow as on the left in Fig. 12.1 (page 162), then at the place where two patches meet there will be a discontinuous change from one direction of symmetry to another. You can often see these places on the surface of a

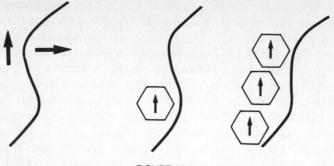

FIGURE 12.1

pond—spots where the ice is slightly thicker and lumpier than elsewhere. The line of suturing is called a "defect" in the crystalline ice.

The point to note about defects is that they involve energy. To see why this is so, imagine what would happen if you brought a single ice crystal up to the edge of one ice patch, shown as a wavy line in the center of Fig. 12.1. The force exerted by the atoms in the patch would turn the crystal so that it would orient itself along the same axis as the patch, which is how the ice normally grows, adding new material in alignment with the existing patch.

Now look at the situation another way: The configuration containing the new crystal aligned with the patch is the state of lowest energy to which the system can go. Left to itself, the system will go down to this state as naturally as a ball left on the side of a hill will roll to the bottom. To keep the crystal in a different alignment, as it would be if it were in a patch on the other side of the defect, energy has to be put into the system. Consequently, when a defect forms and a large number of crystals are kept out of their natural alignment (as shown on the right in Fig. 12.1), there must be energy locked up in the system. If the alignment could be changed, energy could be extracted from the system.

You've probably followed the argument up to this point. Now comes the part where you may balk. Because the defect has energy stored in it, it weighs more than the neighboring ice.

This is a consequence of $E = mc^2$. It is not a consequence we normally think of, but it is true nonetheless. A body to which energy is added has more mass — and therefore weighs more — than it did before. A cocked mousetrap weighs more than an uncocked one. We don't think about such things in workaday life because there's no scale that could measure so small a change; it would be far less than the mass of even the smallest elementary particle. It is only when we deal with very large masses and very large energies under the conditions that obtained in the early universe that we have to remember the full implications of mass-energy equivalence.

If we looked at the surface of a newly formed ice sheet we would see something like what is shown in Fig. 12.2. Against a more or less uniform background corresponding to the ice patches that grew independently of each other, there would be a series of veins of slightly higher mass produced by the defects that formed along the sutures of the patches.

The same sort of thing is supposed to have happened when the universe froze at 10^{-35} second. The universe did not change into its new state suddenly, all at once. Like the ice on the pond, the new state grew out from various nucleation points. Defects like the ice sutures formed and, because of the enormous energies available at that time, acquired large masses. Cosmic strings are one type of defect that could form in the freezing of the universe.

Many, but not all, of the GUT and supersymmetry theories predict the formation of strings in the freezing at 10^{-35} second. And although various theories do not predict identical strings, they do predict stings with the same general properties. For one

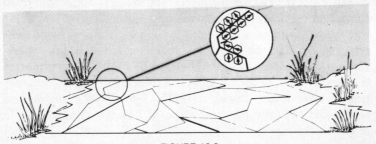

FIGURE 12.2

thing, as we mentioned, the strings are very massive. They are also very thin—the distance across a string is much less than the distance across a proton, for example.* The strings carry no electrical charge, so they do not interact with radiation as ordinary particles do. They come in all shapes—long wavy lines, vibrating loops, three-dimensional spirals, and so on. Clearly, the strings are perfect candidates for dark matter. They exert a gravitational attraction but cannot be broken up by radiation pressure in the early universe.

Strings and Galaxies

Once stable strings are formed, they last for a long time, as we shall see. From the time of their formation at 10^{-35} second, they constituted a massive, lumpy background against which the evolution of particles, nuclei, and atoms played itself out. Since they are not affected by radiation pressure like a plasma, they can serve as the condensation nuclei—the seeds—for the formation of galaxies, galactic clusters, and superclusters, provided that they can survive long enough to do so.

The main spokesman for the idea of cosmic strings is Niel Turok, a young theoretical physicist. Nominally based at Imperial College, London, Turok spends a great deal of time in the United States, including in his itinerary, I am happy to say, my old department at the University of Virginia. He has made his life's work (insofar as that term can be applied to a man not yet in his thirties) the unraveling of the behavior of cosmic strings—the working out of the complex quantum field theory equations that describe them. His approach is admirable in its thoroughness: Instead of following the normal path by working out the behavior of strings and leaving it to others to find the effect strings have on the galaxy problem, Turok and the young men around him have decided to learn cosmology. It is not often that researchers, in the full heat of exploring new territory, take time out to educate themselves in this way. Even more unusual is the fact that the harshest critic of cosmic strings, P. J. E.

* The estimated thickness of a string is 10^{-30} centimeter, compared to 10^{-13} for a proton.

Peebles of Princeton, has been acting as their tutor. This collaboration, to me, is an expression of the highest ideals of the scientific community.

The picture that emerges from Turok's work seems to carry within it the solution to many of the problems posed in this book. It is also easy to visualize. According to the computer simulations, during the GUT freezing, the defects formed a long, continuous chain such as is shown in A of Fig. 12.3. The various segments of the string whipped around through space, moving at almost the speed of light. When the primordial string crossed over itself, as shown in B, Turok's work shows that the exposed loop separates and breaks off.

After a short time, then, the universe was full of bits and pieces of differently shaped cosmic strings. Some were snaky, loosely flopping lines. Such a string will attract the surrounding matter into a plane as it flops back and forth. A commoner shape for a string was a closed loop of the type shown in B. Such loops are what I had in mind when I mentioned the Worm Ourobouros. Around it, the matter being pulled in toward the loop would collect in a spherical or cigar-shaped cluster.

The interesting thing about these shapes, though, is not so much their geometry as their evolution. For example, if, during the course of its vibrations in space, a loop should assume a figure-eight shape when one line of the string crosses over another, then the figure-eight will split up into two separate loops, each a piece of the original "eight," as shown in C of Fig. 12.3. Turok calls this the process of "shedding loops."

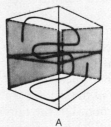 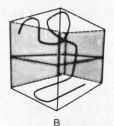 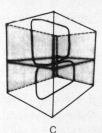

A   B   C

FIGURE 12.3

165

So cosmic strings, once created, do not remain static entities in space. They change as they move around, with large loops becoming smaller loops in the course of time. Although this makes the theory of strings more complicated, it also gives us a chance to test it against observation. After tracing the evolution of the strings in time, we can ask ourselves how the matter that is attracted to them is arranged. If the theory tells us that the strings would all break down into galaxy-sized bits (or smaller) before the decoupling of radiation, then string theory is predicting a universe in which galaxies are scattered more or less at random throughout the void. If, on the other hand, the strings stay the size of galactic superclusters, we expect a universe in which all galaxies are strung out along the lines in space marked out by the strings. The real universe is, of course, somewhere between these two extremes, although it is closer to the latter than to the former.

One of the great triumphs of the work Turok and his collaborators have done is their calculation of the correlation function (see page 72) to be expected for galaxies if the strings really exist. The correlation function, you will recall, is the probability that if a galaxy exists at a particular point in space, there will be another galaxy located within a certain distance of the first. In a completely random universe you would expect to find galaxies separated by almost any distance; no separation would be more likely than any other. In a rigidly ordered, Tinkertoy universe, on the other hand, where all the galaxies are spaced at exactly the same intervals, the correlation function would be zero everywhere but at that chosen separation.

Given the mechanics of the cosmic strings, it is possible to work out how often matter will cluster around various-sized loops and planes, and from this to predict the number of times galaxies can be expected to occur in groups of various sizes. This means that when the expected sizes of the cosmic strings are worked out, the probable separations of galaxies can be worked out as well. In Fig. 12.4 is shown what I believe to be the strongest evidence available for the reality of the strings. On the vertical axis we plot the correlation function. On the horizontal line we plot the distance of separation. The solid line is the prediction of what this probability ought to be from string theory, while the dots represent the data obtained from observations. (The vertical lines attached to the dots indicate the uncertainties astronomers feel are inherent in their measurements.) The agreement between the prediction and the observation is striking.

If we accept this evidence, our picture of the early universe becomes still more striking. Shortly after the GUT freezing, the universe was a giant snakepit in which cosmic strings, both free and in loops, were whizzing around, colliding with each other, shedding loops, and taking part in the universal expansion. As time went on and normal matter went through the various freezings outlined in Chapter 3, the strings stayed in the background, evolving according to their own laws. Throughout this period they also exerted a gravitational attraction on the other matter present, causing it to collect into the lumps that eventually became galaxies. These collections of matter around the strings

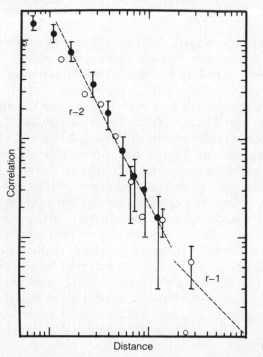

FIGURE 12.4

would consist of *both dark and ordinary matter*, since both are affected by the gravitational force. In this view, the galaxies took form with their complements of luminous and dark matter already in place. The complicated arguments about biasing needed for the cold-dark-matter theories are unneccesary here.

There is, moreover, no difficulty about explaining structures such as the great Pegasus-Perseus supercluster. A long cosmic string will naturally attract matter to it, and this matter will naturally form galaxies that appear to be arranged like beads on a string. Other shapes of strings will form other sorts of clusters. The steadfast refusal of the universe to show homogeneity when we look at the largest scales simply reflects the fact that the universe isn't homogeneous. This character is a legacy of that early moment when the GUT freezing formed strings the way ice forms structures on a freezing pond.

Where Are They Now?

Cosmic strings give us a particularly appealing view of the universe. It seems, for example, that at the core of every galaxy there is a cosmic string, coiled up like the Worm Ourobouros. Did these old mythmakers really come so close to the truth? Could one perform a tightrope walk (in principle, at any rate) along the billion-light-year stretch of the Perseus-Pegasus supercluster? Are the cosmic strings, in other words, still all around us?

Unfortunately, current theories tell us that this somewhat romantic view of the universe is unlikely to be valid. The reason is simple: Cosmic strings do not live forever but slowly die out. The largest strings live longest, the smallest die most quickly. Only the very largest strings could have survived to the present.

The process by which cosmic strings disappear can be understood with the help of a familiar analogy. If you pluck a guitar string, it vibrates and waves are set up in the air. These waves travel from the string to your ear, where they are perceived as sound. The waves consist of moving air molecules, the energy for the molecular motion being supplied by the string. As the vibration continues, the initial supply of energy is slowly drained away by the creation of sound waves. Eventually, the energy is used up and the vibration stops.

So it is with cosmic strings. In the initial "snakepit" stage of the universe the strings were all set into violent motion; they were, in effect, "plucked." The general theory of relativity tells us that when something as massive as a cosmic string is accelerated, it will give off waves. It won't give off sound, of course, since there's no air in intergalactic space. It won't give off light, either, or any other kind of electromagnetic radiation, because the string carries no electrical charge. What comes off is something called gravitational waves. Just as a sound wave moving by a point causes the air to move, and an electromagnetic wave causes an electrical charge to move, a gravitational wave moving by a point causes matter to move. Whether gravitational waves have actually been detected in the laboratory remains a subject of debate among experimental physicists, but for our purpose we need only note that they provide a way for the cosmic string to radiate — to convert its energy into waves.

But there is a difference between the guitar and a cosmic string. For the guitar, the energy available for conversion into sound is the energy of vibration alone. When it is used up, the string stops vibrating and just sits there. There is no mechanism (short of putting the string into the core of a nuclear reactor) by which the mass energy of the string can be converted into sound as well.

A cosmic string, on the other hand, is all energy to begin with. When it starts to give out gravitational waves, the process goes on until the string has simply radiated itself out of existence. When its energy is gone, there is nothing left. It should therefore be possible to use the rates of energy loss predicted by the general theory of relativity to calculate how long the energy stored in any cosmic string will last.

As a matter of fact, there was a nervous period in the spring and summer of 1986 when it appeared that cosmic strings would have too short a lifetime to do their job in the formation of galaxies, that they would shed loops and radiate themselves out of existence before matter and ordinary radiation decoupled. The calculations now seem to show that loops capable of forming the seeds of galaxies would last long enough to carry out this function, but would not have survived into the present. In other words, there is no Worm Ourobouros at the heart of the Milky Way, although the strings that give rise to galactic clusters and superclusters are still around.

According to the current version of the theory, the Milky Way originally condensed around a string with a mass of about a hundredth the mass of the present galaxy and about one hundred light-years long. In other words, the galactic "seed" weighed in at a mass of about 100 million suns and a length comparable to thirty normal star separations in today's galaxy. After it had collected *both* the luminous and dark matter of the Milky Way, it radiated itself out of existence, probably disappearing in a puff of "smoke" made up of the kinds of exotic particles discussed in Chapter 11. Its only legacy: the Milky Way galaxy itself.

Searching for the Strings

You may find this scenario a little too slick. That the vexing problem of galaxy formation should be solved by a theoretical construct like the cosmic strings, and that strings should have obligingly disappeared so that they are undetectable today, may well seem a little too good to be true. It has some of the glibness of a television commercial. Some prominent cosmologists (in private, at least) call strings "snake oil," referring to the old-time medicine salesman who used to tout bottles of elixir that was "good for what ails you." In informal sessions over late-night pitchers of beer, I have also heard the tooth fairy mentioned in this context. Skepticism is probably to be expected, and those who make these comments are not being harder on the theories of others than they are on their own. But things are perhaps not so bad as the doubters make them seem, for it is possible to search for independent evidence of the existence of strings in the universe today.

There are two ways to find such evidence. One, the so-called gravitational lens, relies on the effects that cosmic strings would have on light from distant galaxies. The other method, somewhat more indirect, involves a search for the gravitational waves given off by the string early in the life of the universe.

A gravitational lens is shown in Fig. 12.5. Light from a distant luminous object comes to the earth past an intervening, very massive, body such as a distant galaxy or a string.

You may recall that according to the general theory of relativity, light from a distant star is bent as it passes near the sun or another massive body. In the situation shown in the figure, the intervening body will bend the light rays as shown. Someone standing on the earth will therefore see two images of the distant object: one produced by the ray that went over the top, the other by the ray bent under the bottom.

The first gravitational lens (in which a dim but otherwise normal galaxy provided the mass needed to bend the light) was discovered in 1978; today there are roughly half a dozen known. In all cases, the intervening body is luminous and can be identified, although we needn't be able to identify the body that is bending the light in order to have a lens. A massive (and unseen) string would produce the same double image in the sky. A lens without an identifiable intervening body would, therefore, be possible evidence for a cosmic string. A long string might even produce a line of double images—a clear signal.

In the spring of 1986 there was a brief flurry of excitement

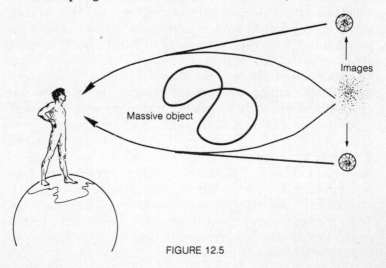

FIGURE 12.5

171

when astronomers at Princeton University announced the discovery of what they thought was a double image of a quasar in the constellation Leo (the Lion). Unfortunately, subsequent work showed that the two quasars were not a twin image, but actually two different bodies. At the moment, therefore, there is no observational evidence for a gravitational lens that is caused by cosmic strings, but the search continues. It's much too early to say anything about what might eventually be found.

The second area in which evidence for cosmic strings might turn up involves pulsars. The end product of the death of large stars, pulsars are rapidly rotating objects only a few miles in diameter. Radio beams are emitted from hot spots on the pulsar's surface, and the rotation carries these beams past the earth, much in the manner of the beam from a lighthouse beacon. Each time the beam passes, we see a pulse of radio waves, and the pulses repeat periodically. The repetition rate is high — anywhere from a few to several hundred per second. Consequently, the pulsars can be thought of as very accurate "clocks" in the sky, ticking away at a steady rate.

If the cosmic strings emitted a great deal of gravitational radiation in the early days of the universe, then most of this radiation must still be present today. Pulsars would therefore find themselves in a sea of gravitational radiation, continuously being jostled and bumped as the waves went by. This, in turn, would impose a small random irregularity on the timing of the pulses as seen from earth. According to calculations now being done, we should be able to see evidence for gravitational waves when our determination of the period of rapid pulsars is about ten times as accurate as at present. Observers feel that this sort of improvement should be possible within the next few years.

So although the disappearance of cosmic strings through gravitational radiation may seem contrived, there are ways of testing the theory by observation. Cosmic strings are not the snake oil they seem to be at first glance. If after thorough searches of the type outlined above we still have no hard evidence for the existence of a cosmic string, we may have to rethink this statement. At the moment, however, it seems to many scientists that cosmic strings provide the best chance of solving the problem of the large-scale structure of the universe.

THIRTEEN

Experimental Searches
for Dark Matter

Heard melodies are sweet, but those unheard
Are sweeter. . . .

— JOHN KEATS,
"Ode on a Grecian Urn"

I T SEEMS OBVIOUS that the question of the nature of dark
matter is not going to be laid to rest until someone actually
gets hold of a piece of it in a laboratory. It's all very well to
produce theories and show that dark matter must behave in this
or that way, but until we can isolate some of the stuff and
actually *see* it behaving as it's supposed to, a great many people
(myself included) are going to be unsatisfied. Even for invisible
dark matter, seeing is believing.

Two roads can be traveled on the way to finding experimental
evidence for the existence of the particles that are supposed to
make up over 90 percent of the mass of the universe. One is to
try to produce these particles in accelerator laboratories, the
other is to build instruments that will detect them as they sweep
by the earth. Both methods are being vigorously pursued at the
present time. What follows is a description of some typical

experiments that either have been completed or are being planned for the next decade.

A two-pronged attack is in keeping with the history of particle physics in our time. Many of the elementary particles whose existence we now take for granted first burst upon our consciousness in reactions initiated by cosmic rays — particles raining down on the earth from supernovae in the Milky Way. At the same time, many of the most important discoveries came as the result (either by design or serendipity) of experiments at major accelerator centers.*

Experiments at Accelerators

An accelerator is a device that produces a beam of particles — either protons or electrons — that are traveling at velocities near the speed of light. These particles are directed at a target — almost any collection of atoms will do. In some of the resulting collisions, part of the energy of the beam will be converted $(E = mc^2)$ into the mass of new particles. No matter how unlikely it is that a particular particle will be produced in such a reaction, if we have enough energy in the beam and wait long enough, sooner or later we'll see what we're waiting for. The hope of present experimenters is that this will be as true in the search for dark matter as it has been in the search for other particles in the past.

Before we look at particular experiments, we should note a few points. First, the energy that can be imparted to any beam by an accelerator is necessarily finite. For any machine, in other words, there is an upper limit to the amount of energy available for conversion into mass. This means that a negative result in a search can never be taken as conclusive evidence against the existence of a certain form of dark matter, but only as a statement that the particle being searched for has a mass greater than the greatest that could be produced by the energy of that particular machine. It is always possible that the next higher-

---

* The history of the development of particle physics, as well as the design and use of accelerators, is discussed in my book *From Atoms to Quarks* (Scribners, 1980).

energy machine built will produce in abundance a particle that cannot be seen today.

The second point is that many of the more exotic candidates for dark matter discussed in Chapter 11 cannot be produced singly, but must always be created in pairs. The theories tell us, for example, that we cannot produce a single, isolated photino in any reaction initiated by the impact of an electron or proton on ordinary matter. What must happen is the production of a pair of photinos. Similarly, we cannot produce a single selectron, but must produce a pair — a selectron and an antiselectron. In effect, this cuts in half the energy available for conversion into mass in any accelerator, for the energy must be shared between both members of the pair.

The best design for a machine to search for new particles requires what physicists call a colliding-beam apparatus. A typical machine of this type is sketched in Fig. 13.1. An accelerator produces beams that are injected into large rings, where powerful magnets keep the particles circulating. Particles with a positive electrical charge (such as protons) move in one direction around the ring, while particles with a negative electrical charge (for example, antiprotons) move in the other direction. The ring is designed so that at certain places, such as those marked with

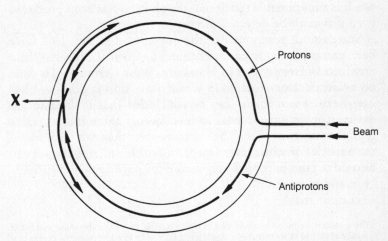

FIGURE 13.1

an X in the figure, the two beams collide head on. At that moment, all of the energy imparted to the projectiles is available for conversion into mass—there is no more efficient system possible. Consequently, it is with machines of this type that the main searches for dark matter have been carried out.

One such search was carried out at a number of machines that create high-energy beams of electrons and positrons.* The idea is that when these particles come together we can get reactions such as that shown in Fig. 13.2. An electron and a positron produced by the accelerator collide, creating (for example) their opposite numbers in the supersymmetric world, a selectron and an antiselectron. In Chapter 11, we saw that supersymmetric particles would always give off energy until they had become the lightest particle possible, and in this case it would mean that the selectron and antiselectron would eventually turn themselves into photinos and ordinary particles, as shown.

As I pointed out, it is impossible to detect the photinos directly, because their interaction with ordinary matter is so weak. We can, however, detect the electron and positron that result from the decay of the selectron and antiselectron. There are characteristics of the resulting electron-positron pair that would arise from a process like that shown in the figure but would not be seen in pairs produced by any process involving normal matter. It is thus possible to tell that the photino has been produced even if it can't be detected directly.

Searches for reactions like the one shown in Fig. 13.2 have been carried out at research institutes at Cornell University and Stanford University and in Hamburg, West Germany. To date, no evidence for a reaction in which a photino is produced has been seen. From this it can be concluded that if the photino exists, it must have a mass at least twenty-three times greater than that of the proton. The experimenters may not have found the particle for which they searched, but this inferred result can be used to put limits on the properties the particle may eventually be found to have.

---

* The positron is the antiparticle of the electron. The two have the same mass, but opposite electrical charge. If they collide, all of their mass is converted into energy and the two particles annihilate themselves.

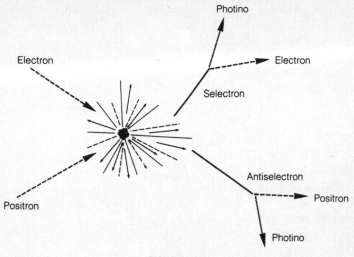

FIGURE 13.2

A similar search for supersymmetric particles can be carried out at machines that produce colliding beams of protons and antiprotons, such as the great accelerator run by the European Center for Nuclear Research (CERN) in Geneva, Switzerland. At that machine, reactions such as the one shown in Fig. 13.3 (page 178) can, in principle, be created by head-on collisions between the proton and its antiparticle.

As was the case for the experiments carried out at electron machines, the photinos produced in this other sort of reaction cannot be detected directly. What can be detected is the missing energy that these two particles would necessarily carry off. You can get some idea of this process by imagining that you are riding along on the squark in the upper branch of the reaction shown in Fig. 13.3. When the squark decays into a photino and a spray of normal matter, you have a situation such as that shown on the left in Fig. 13.4. The normal matter would come off in a jet in one direction, balancing the photino, which would move off in the other.

If the squark were surrounded by a detector, as shown in the figure, then the jet of normal particles would be detected, but not the photino. We would see an unbalanced situation in which

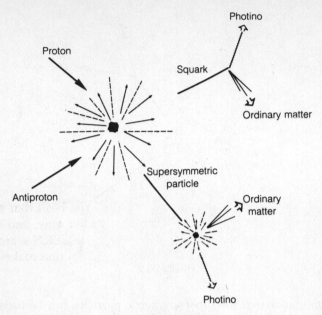

FIGURE 13.3

particles move off to the right with no countervailing particles moving to the left. If you imagine rolling the detector out from a sphere onto a flat plane, you would get a reading like the one shown on the right—a single spike (corresponding to the jet) against a low background.

Nobel laureate Carlo Rubbia, working at the UA-1 area at CERN,* has indeed seen some events of the type shown. The sketch of the detector output on the right in Fig. 13.4 is adapted from some of his data. As of the spring of 1988, physicists do not interpret these results as evidence for supersymmetric particles. It is possible to get signals like the one shown for interactions involving only normal matter. For example, the undetected particle could be a neutrino or something else that does not readily register in a laboratory instrument. The real question is whether the events seen at CERN occur more often than one

---

*UA-1 stands for "underground area 1." CERN stands for "European Center for Nuclear Research," a laboratory in Geneva, Switzerland.

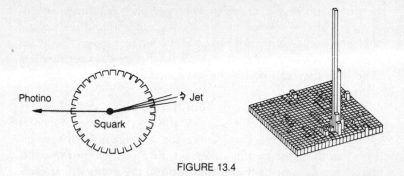

Photino

Squark

Jet

FIGURE 13.4

would expect them to on the basis of the hypothesis that they are caused by the interactions of normal matter only, and that does not seem to be the case. Nevertheless, the CERN searches show one way to try to detect the dark material that makes up most of the universe.

"Cosmic Ray" Dark Matter

A new type of detector being tested by Blas Cabrera and his co-workers at Stanford University may provide a way of detecting dark matter directly, without having to create it in an accelerator. To understand what is involved in this venture, you must know two facts, one involving the structure of the galaxy, the other involving modern high technology.

The first fact is that, as we saw in Chapter 6, the luminous matter in the Milky Way is embedded in a sphere of dark matter. The dark matter does not rotate at the same rate as the galaxy, so the earth can be thought of as moving through a "wind" of dark matter. In the same way, when you drive your car on a still day, you feel a wind created by the movement of your car through the air. If we had a sensitive enough instrument, we ought to be able to detect this dark-matter wind. At least, this is the hope of people who are working on the project.

The second fact is that thanks to the growth of the micro-electronics industry over the past several decades, it is now possible to grow large crystals of incredibly pure silicon. Indeed, the little chip that operates in your personal computer or calculator probably started life as part of a single cylindrical crystal of silicon some six inches across and four feet long. The availability of such large, perfect crystals is what gives experimenters hope that they can detect the dark matter through which the earth is moving.

The principle of a detector built from a silicon crystal is very simple. Visualize the atoms inside the crystal as tied together by a complex, interlocking, Tinkertoy arrangement of springs. If one atom in the crystal collides with a particle in the dark-matter wind, some of the springs around the atom will stretch. After the collision, the atom which was struck will be jiggling, and this vibration will be transmitted from one atom to the next by the network of springs. Eventually, the disturbance will work its way from the interior of the crystal to the surface. If we have sensitive enough detectors there, we should be able to tell that an interaction has occurred, and by measuring the time of arrival at the various surfaces, we can tell where within the crystal it happened.

It is easy to describe this procedure; making it work in a laboratory setting is another matter. Any imperfection within the silicon crystal will change the way the "springs" are arranged and, in effect, smear out the signal as it travels to the surface of the cube. That is why this sort of detector is being developed only now, after industrial scientists have learned how to grow crystals that are both large and perfect.

Cabrera and his team plan to instrument a series of one-kilogram blocks of silicon with a grid of detectors fastened to each side, and then cool the entire system to within a few degrees of absolute zero. The cooling not only cuts down random motions within the silicon, making the signals at the surface cleaner, but also allows the experimenters to use extremely sensitive superconducting detectors to measure the signal. Their plan is to stack up a thousand such cubes and see what happens. This exercise in high technology is not a huge affair — a ton of instrumented silicon would fit comfortably under the average kitchen table.

There are several interesting points to make about this experiment. The kind of vibrations set up in the silicon by collisions are very similar to the vibrations that characterize sound waves in air. Thus the detection technique is similar to finding a tennis ball by listening when it bounces against a wall. You could say that Cabrera plans to *listen* for dark matter.

Because interactions with dark matter are expected to be very rare, it will not be possible to turn the detector on and just wait for an interaction. There are other particles around that will produce signals, no matter how well shielded the system is. Indeed, the detector's first job will be to detect solar neutrinos (see p. 157). To detect dark matter, the experimenters will have to be clever. Their plan is as follows.

We know that the earth orbits around the sun once a year. This means that for half the year the earth must be heading into the dark-matter wind, while for the other half it must have the wind at its back. Consequently, more dark matter will flow through the apparatus for one half of the year than for the other. It is this annual signal, sitting on top of everything else, that the experimenters hope to find.

That Cabrera should be contemplating a search for the dark-matter wind today is intriguing, because a century ago two other American physicists executed a now famous search for another sort of "wind." Working in Cleveland, in what eventually became Case Western Reserve University, Albert Michelson and Edward Morley built an experiment similar in plan to the dark-matter detector just described. At that time, scientists believed that the universe was permeated by a substance known as the "ether," and that this substance provided a kind of framework for the entire universe. The ether defined the "God's eye" frame of reference that was implicit in Newtonian physics and that was explicitly rejected by Einstein. The motion of the earth through the ether was supposed to produce an ether "wind," and this wind would have to appear to change direction from one part of the year to the next because of a process similar to that outlined above for the dark-matter wind. When Michelson and Morley showed that such an ether wind did not exist, they removed one of the conceptual barriers in the way of the eventual development of the theory of relativity.

Michelson and Morley started out looking for an effect and wound up finding nothing. Cabrera, on the other hand, is starting out with the hope that he won't see a change in the dark-matter "wind" as the earth moves around its orbit. "After all," he says, "even if we find something, we'll have no idea what it is."

FOURTEEN

The Fate of the Universe

Tomorrow, and tomorrow, and tomorrow. . . .
— WILLIAM SHAKESPEARE
Macbeth, Act V, Sc. 5

A Summing Up

THE MOST STRIKING FACT about the way things are moving in modern cosmology is that the loose ends seem to be coming together. It used to be that one could put forward a completely wild idea and be sure that there was enough ignorance about the universe to prevent any embarrassing conflict between the idea and what was known. As on old maps, many places in the cosmos were marked "Here Be Monsters." This is no longer true. For example, we cannot make free and easy assumptions about the nature of dark matter. We know enough about the way the nuclei of atoms were formed three minutes into the Big Bang to put severe limits on how much of that matter can be in the form of baryons. We know enough about large-scale structure so that it is no longer enough to show that a particular assumption about dark matter will solve the galaxy problem; we must also show that it explains the voids as well. Indeed, we have reached the point where the creation, evolution, and present structure of the universe appear as a single, seamless problem. It is no longer possible to deal with just one piece of the puzzle—we must solve the whole of it at one stroke. Theorists have to show that in settling the question of structure, they do not mess up the agreement between theory and observation that has been obtained for the early stages of the Big Bang—that they will not produce galaxies at the cost of eliminating the voids.

An incident I witnessed at a recent international conference will make this point clear. A famous cosmologist got up after a talk and suggested a mechanism that might allow all the dark matter to be in the form of baryons and still produce roughly the same abundance of helium we see in the stars (the connection between these two is discussed in Chapter 9). No sooner had he sat down than a young theorist rose in the back of the room and pointed out that if this mechanism were adopted, the helium might come out right, but the abundance of lithium would be wrong. The senior man withdrew his remark.

The point of this anecdote is that ten years ago this interchange could not have taken place. Then we lacked sound observations on such things as the abundance of lithium in the universe and our calculations had not progressed to the point where we could make unambiguous predictions about what that abundance ought to be. Consequently, imagination could be given full sway, and nobody could put definite restraints on it.

We are beginning to know enough about the universe to narrow our options considerably. Of all the universes we can build in our minds, fewer and fewer are surviving the twin tests of observation and calculation. The time is coming when the question asked in cosmology—"How did the universe come to be the way it is?"—will be so bounded on all sides by firm knowledge that we may even have trouble finding *any* solution that doesn't cause discrepancies somewhere. I won't be at all surprised if in the next few years the difficulties theorists have in finding a solution to the problem of evolution and structure give rise to a new philosophical movement. It would claim that the rational, reductionist approach of Western science has reached the end of its road and some new (preferably mystical) line of attack must be tried. This happened in the 1970s when particle physics hit a temporary impasse, and it could happen again in cosmology. But just as the development of the unified field theories broke the particle-physics logjam in the 1970s, I expect that, should this sort of problem arise in cosmology, the tried-and-true methods of theoretical science will eventually be able to deal with it.

For the more closely you look at the new map of the universe being built today, the more you realize that what it discloses is a universe forming a single, marvelous machine; all the gears and engines fit together beautifully to make a coherent whole. Everything fits together, and our realization of this fact may well be the most important insight to come out of the new cosmology.

A Few Far-Out Ideas

Having made the point that there are many firm new constraints on our models of the universe, I must hasten to add that even so, there is still a fair amount of room for imagination. By way of illustration, let me mention a few lines of attack that at this moment have not been shown to be wrong.

The cosmic strings discussed in Chapter 12 arose from the way the strong force froze out from the others when the universe was 10^{-35} second old. They arise, in other words, from the grand unified theories. Recently, Ed Witten and Jerome Ostriker at Princeton have pointed out that cosmic strings can also arise in supersymmetric theories, where gravity is fully unified with the other forces. In some cases, these supersymmetric strings have unusual properties, to say the least.

The most unusual is that when a normal particle comes near a string it can fall in, giving up its energy as it does so. Once inside the string, the particle is trapped, and remains inside until some outside agency adds enough energy to free it. These trapped particles can move, and if they are things like ordinary electrons, this motion constitutes an electric current. Calculations indicate that these trapped particles can produce enormous currents without loss — quadrillions of times as intense as that carried by the biggest power line.

If there is a magnetic field in the early universe, then the movement of the strings through it will push these current levels up to the point where the string explodes, releasing all of its particles in a massive rush. The suggestion is that it was the explosions of these supersymmetric strings early in the life of the universe that created the voids we see today.

This ingenious theory suffers from the defects of all explosion theories (see Chapter 6), but in addition it leaves unanswered what is perhaps the most interesting question of all — where did the original magnetic field that generated the current come from? Until these questions are faced, it will be hard to think of this version of cosmic strings as anything more than an interesting hypothesis.

Another speculation, one that probably stands a fair chance of success, is being pursued by Neil Turok and David Schramm at the University of Chicago and the Fermi National Accelerator Laboratory. Recognizing that every type of dark matter, taken by itself, has difficulties in explaining all of the other conditions we have ascertained about the universe, these theories adopt an "all of the above" position. For example, we saw in Chapter 10 that hot dark matter in the form of massive neutrinos cannot explain how galaxies could have formed. But what if there were two kinds of dark matter in the universe — neutrinos *and* cosmic strings? The strings do very well at explaining galaxy formation and the neutrinos do well for large structures. Why not take them together and see if the strength of one doesn't wipe out the weakness of the other?

It's still too early to say whether this suggestion will work out, but there is every reason to believe that it will. If it does, then perhaps the kinds of exotic dark-matter candidates we talked about in Chapter 11 won't be necessary. It would be pleasant to be able to go no further afield than the particles with which we have been long acquainted.

What Would It Taste Like?

One of the great joys of teaching undergraduates is that every once in a while one of them asks you something that opens vistas one would never have thought of oneself. Last year, tired of reading the same old term papers in my introductory physics class, I tried an experiment. I told the students to read and report on five articles of their choice in science magazines intended for the general public. I wanted to get them used to the idea of getting information about science for themselves, outside of a university classroom.

One of the students read an excellent article on dark matter written by Lawrence Krauss in *Scientific American*. After producing the usual report, the student made a comment. It said, in effect, "This is all very good, but what dark matter does to the universe isn't going to affect my life very much. What I want to know is more personal — what does dark matter taste like? does it feel slimy? could I swim in it?"

This made me stop and think. Like most physicists, I took it for granted that dark matter had to do with galaxies and superclusters, not with everyday experience. Yet if it really exists, it should be possible to put together enough of it to taste or jump into. What would the experience be like?

To answer such a question, you have to think about what it means to taste or feel something. Taste involves a chemical reaction in which molecules of a substance combine with molecules in taste buds to produce electrical signals that go to the brain. Feeling involves the stimulation by pressure of specialized receptors in the skin. Thus if we are to taste dark matter it will have to be capable of forming itself into atoms or molecules, and if we are to feel it, it must be cohesive enough to exert a pressure.

By this test, we can immediately rule out dark-matter candidates like neutrinos and axions as far as taste goes. They don't form atoms, and their interactions with normal matter are so meager that they would make no impression whatsoever on taste buds. In fact, we have all been "tasting" neutrinos all our lives in the sense that they've been passing through our mouths at the rate of many millions per second in their trip away from the sun. They don't trigger any of the necessary chemical reactions as they pass. The same would be true of axions (if they exist). We have been "swimming" in neutrinos all our lives as well, but they are no more capable of exerting pressure than they are of tickling the taste buds.

Nor could one taste shadow matter, for the simple reason that, by hypothesis, it doesn't interact chemically with ordinary matter. You might think that because shadow matter can clump together into solids and liquids it could be felt, but that's not true. Because shadow matter interacts with us *only* through the force of gravity, if someone put a lump of it in your hand, the lump would fall right through, causing almost no disturbance to the tissues. This is so because when you hold something in your hand it is the electrical forces between the atoms in your hand and the atoms in the object that overcome the force of gravity and keep the object from falling. That force cannot exist between shadow matter and your hand, so the shadow matter would have nothing to hold it up.

The same thing would probably happen if you tried to grab a piece of cosmic string, but for a different reason. You might suppose that a string is so massive that if you tried to hold a loop in your hand it would fall through immediately, leaving behind a hole to mark its passage like some sort of cosmic cookie cutter. In point of fact, even the great mass of the string cannot overcome the inherent weakness of the gravitational force. As the string fell through your hand, the force it would exert on the average atom would be far less than the ordinary electrical forces exerted by neighboring atoms — the forces that hold the tissues in your hand together. The string would fall through your hand all right, but its effects on the atoms it encountered would be too small to dislodge them. You wouldn't feel a thing.

So our search for tasty dark matter comes down to supersymmetric partners. The most "common" particle — the photino — doesn't form atoms, although it might be capable of exerting a pressure somewhat lower than that of ordinary light. Neither photons nor photinos in ordinary densities could exert enough pressure for us to feel them. The same is probably true of "satoms" made up totally of supersymmetric particles.

If, however, the selectron is stable — a hypothesis contrary to current thinking—there is a possibility of something interesting. This possibility arises because a stable selectron, which has a negative electrical charge, could replace one or more electrons in an ordinary atom. Because of its large mass, the orbit of the selectron would be different from that of the electron it had replaced. Consequently, all of the other electron orbits in the atom would be thrown out of kilter and all the chemical properties of the material in which it resided would be changed. Such "super-atoms," therefore, would have a totally new set of chemical reactions. Foods that had "super carbon" in them would not taste like any food ever known.

So if supermatter exists, it may open up new culinary frontiers. And who knows—we may be able to develop something that tastes like ice cream but has no calories!

The Fate of the Universe

What effect will dark matter have on the ultimate fate of the universe? Several things can be said about this question. In the first place, as we shall see shortly, it will have almost no effect on the future as seen by an observer on the earth. But if the universe really does possess the critical amount of matter, then the existence of dark matter will indeed have an effect on the long-term future. It used to be customary in discussions of this sort to entertain the idea that the universe was cyclical—that the Big Bang would be followed by a collapse (the Big Crunch) and another expansion (the Big Bounce). But if our current ideas are true, this will not happen. The universe has one shot at existence—one explosion followed by an expansion that slows down for an infinite length of time.

We can follow the course of the universe under the assumption that the laws of nature we now observe will always hold true in the future. From the standpoint of an observer on the surface of the earth, the large-scale structure of the universe makes very little difference to the appearance of the night sky, since distant galaxies are, by and large, invisible to the naked eye. The stars in the Milky Way (including the sun) will continue to burn until they use up their store of hydrogen and helium fuel. In the sun, the fuel will run out in about four billion years, at which point it will evolve into a red giant, a swollen star whose orbit extends past the orbit of Venus. To an observer on earth, the sun will appear to fill half the sky. At that time, the oceans will boil and any life left here will perish. If the human race hasn't had the sense to colonize the stars, this is the end.

Following its red giant stage, the sun will collapse into a white dwarf—a star about the size of the earth that slowly cools off, a cosmic cinder that has lost its source of fire. The stars in the sky will go out one by one, either with a spectacular explosion or with a whimper, like the sun. Should there be an observer on our planet when the universe is a quadrillion years old (a thousand times its present age), the sky would be dark indeed. Almost all the stars we now see would either be so dim as to be invisible or appear as faint points in a sea of blackness. Distant galaxies, never an important part of the nighttime display, would also be diminished.

The slow cooling of the stellar cinders would go on for a long time, the only relief being the falling of stars and gas into the black hole we believe to be at the center of the Milky Way. Occasionally a particle and an antiparticle would find and annihilate each other, adding to the expanding sea of radiation. The universal expansion would continue, but the rate would slow perceptibly as the ages passed.

There will be only two milestones to mark changes as time goes on. When the universe is around 10^{36} years old—long after all the stars have stopped shining—the protons in normal matter will decay. Anything left around in the form of stellar cinders or lumps of rocks will disappear in a puff of radiation as its atoms fall apart. The earth will disappear beneath our feet. Then, when the universe approaches 10^{65} years of age, the black holes that have been collecting matter until this time will start to radiate away their mass in the form of normal energy. They too will die. After this has happened, there will be nothing left in the universe made up of normal matter but a cold, expanding sea of radiation interspersed with a few odd particles that have somehow escaped annihilation and are now too thinly spread to meet any fellow particles again.

Although no physicist has, to my knowledge, considered what

might be happening to dark matter while this story plays itself out, I expect that the dark matter would be going through a similar kind of evolution. Perhaps over the ages we might see the galactic halos collapse into disks, but these disks would be made of photinos and could not form "superstars" or any other sort of interesting structure. Eventually, the photinos would have to fall into their own versions of black holes, which would then radiate themselves away.

So no matter what the composition of the universe, the end will be the same — a cold, expanding sea of radiation from which all life has long since vanished.

A Conclusion

Confronted with this sort of scenario for the end of the universe, both scientists and poets seem to be at a loss for words. Nobel laureate Steven Weinberg, a man who is as responsible as any other for our current understanding of nature, closed his marvelous book *The First Three Minutes* with the gloomy comment "The more the universe seems comprehensible, the more it also seems pointless." Almost a century earlier, the Victorian poet Algernon Swinburne expressed a kindred idea in "The Garden of Proserpine":

> From too much love of living,
>   From hope and fear set free,
> We thank with brief thanksgiving
>   Whatever gods there be
> That no man lives forever;
>   That dead men rise up never;
> That even the weariest river
>   Winds somewhere safe to sea.

And he was only dealing with the second law of thermodynamics, not the Hubble expansion!

There's no question about it — contemplating the end of the universe seems to bring out the gloomiest side of both scientists and poets. The prospect reminds me of a science fiction story that made a powerful impression on me as a teenager. It was a time-travel story. As characters from different epochs interacted with each other, one of them appeared continuously in the background. A mysterious figure in a monk's cloak, he never spoke until the story reached its climax. At that point, he declared, "I am the last human being. Remember," he intoned, "no matter what you do, no matter how hard you strive, it will all end with me."

It's hard to imagine anything more calculated to inflame the adolescent imagination. Even today, with the advent of years and (I hope) judgment, when I know that the story is bad biology and bad physics, I can feel the power of the literary image. Doesn't everything we've learned about the structure of the universe, about unified field theories and dark matter, merely serve to bolster this sort of fatalistic view of the future? If, billions of years in the future, there is to be no life, no intelligence, no memory of the struggles of humanity, what point is there to existence?

As a scientist and a human being, I have had to wrestle with this question. It is just possible that my resolution of it will help you as you face it yourself. After a long period of indecision, I finally realized that the entire issue can be brought down to a simple problem—how will I act tomorrow? Given what I know about the future of the universe, how will I handle the everyday decisions that make up my life? What I finally came to see was this: It may be true that in a quadrillion years the universe will be a cold, expanding sea of radiation. There may be no one to know how I behave tomorrow, no one to remember what any of us did. But this is irrelevant. The point is that *I* will know tomorrow what I have done, *I* will know whether I was the best person I could be.

And in the end, my friends, that is all that matters.

Index

Institute for Experimental and
Theoretical Physics (Moscow),
137, 139, 140, 142
isothermal model, 62, 63

Jeans, Sir James, 60, 61

Kant, Immanuel, 27
Kepler, Johannes, 88–90
Kipling, Rudyard, 49
Kolb, Rocky, 153
Krauss, Lawrence, 187

Leavitt, Henrietta Swan, 30, 33
lens, gravitational, 170–72
lepton, 45–47, 52, 134, 147
lithium-7, 124, 184
loops, shedding, 165, 169
Los Alamos, 143
Luther, Martin, 17

magnetism, 46
mass, critical, 119, 120, 125
mass concentration, 99, 100, 104
matter, 58–62, 64, 65, 83, 84, 90–
94, 96, 102–4, 107, 108, 115–
17, 122, 129, 168, 170, 177,
178, 179, 187
matter, baryonic, 122–25, 127, 128
matter, cold dark, 98, 101–3, 143,
146, 168
matter, dark, 28, 83, 84, 90–98,
104–7, 116, 117, 121, 124,
128, 141, 144–47, 149, 153–
58, 160, 168, 170, 173, 175,
179–84, 186–92
matter, hot dark, 98–101, 129, 186
meson, mu, 134
meson, tau, 134
Michelson, Albert, 181, 182
Michigan, University of, 77
Middle Ages, 15, 17–20, 24
Milky Way, 23–33, 84, 85, 92,
101, 189, 190
Million Galaxy Map, 71
missing mass, 105, 116, 117, 129

mixing, 133, 136
mode, fundamental, 150, 157
molecule, 58, 59, 93
Morley, Edward, 181, 182
Mount Wilson Observatory, 29, 33,
40
multidimensionality, 152
Museum of Alexandria, 10

NASA, 107
National Academy of Science, 33
nebulae, 26–29, 31, 33, 34
nebulae, spiral, 32, 33
neutrino, 98–101, 128–35, 137–
44, 157, 178, 181, 186–88
neutrino, massive, 134–36, 141–44
neutrino, mu, 135
neutrino, tau, 135
neutrino, solar, 181
neutron, 45, 48, 49, 112, 124, 128,
134, 137, 139, 140, 147, 148
Newton, Isaac, 109, 111, 113–15,
181
Nicholas of Cusa, 19, 40
nucleosynthesis, 49, 52
nucleus, 44, 45, 48–50, 122–25,
147, 164, 183

Of Learned Ignorance, 19
oscillations, solar, 157
Ostriker, Jerome, 185
overtone, 150

parallax, 13, 15, 18, 19
parallax, stellar, 15, 19
particles, elementary, 45, 48, 97–
101, 122, 129, 132, 133, 135,
137, 139, 140, 146–49, 151,
153, 155, 159, 163, 164, 173–
78, 181, 186, 190
particles, supersymmetric, 175–78
Particle, Weakly Interacting Massive
(WIMP), 146, 154, 156, 157
Peccei, Roberto, 155
Peebles, P. J. E., 56, 70, 71, 73, 165
Penzias, Arno, 51, 125

About the Author

James Trefil is the Clarence J. Robinson Professor of Physics at George Mason University in Fairfax, Virginia, and a fellow of the American Physical Society. He is also an officer and founding member of the Society for Scientific Exploration, and the author of two physics textbooks and more than a hundred articles for professional journals. Winner of the AAAS-Westinghouse Award for science writing, he has been a regular contributor to *Smithsonian* and *Science Digest*. His other books include the highly acclaimed *The Moment of Creation, From Atoms to Quarks, The Unexpected Vista, Are We Alone?*, and his Natural Philosopher Trilogy: *Meditations at 10,000 Feet, A Scientist at the Seashore*, and *Meditations at Sunset*. He lives near the Blue Ridge Mountains in Virginia.